Science, Agriculture and Research

Tables and Figures

TABLES

FIGURES

List of Acronyms and Abbreviations

AADP   Ayangba Agicultural Development Project
ACARD   Advisory Council for Applied Research and Development
ACCAR   Advisory Council on Colonial Agricultural Research
ADP   Agricultural Development Programme
ADP   Agricultural Development Project
AEA   agro-ecosystem analysis
AERDD   Agricultural Extension and Rural Development Department
AFRC   Agriculture and Food Research Council
AR   Action Research
AR   Adaptive Research
ARC   Agricultural Research Council
ARI   Agricultural Research Institute
BBSRC   Biotechnology and Biological Sciences Research Council
BHC   benzene hexachloride
BSE   bovine spongiform encephalitis
CAP   Common Agricultural Policy
CF   contract farmer
CG   consultative group
CGIAR   Consultative Group for International Agricultural Research
CIAT   Centro Internacional de Agricultura Tropical
CIFOR   Centre for International Forestry Research
CIMMYT   Centro Internacional de Majoramiento de Maz y Trigo
CIP   Centro Internacional de la Papa
DAO   Divisional Agricultural Officer
DDT   dichlorodiphenyltrichloroethane
DEFRA   Department for Environment, Food and Rural Affairs
            (formerly MAFF)
DFID   Department for International Development
DNA   deoxyribonucleic acid
DTI   Department of Trade and Industry
EA   extension agents
EC   European Commission
ECU   European Currency Unit
EEC   European Economic Community
EU   European Union

FAO	Food and Agriculture Organization
FP5	Framework Programme 5
FSR	Farming Systems Research
FSRE	Farming Systems Research and Extension
GATT	General Agreement on Tariffs and Trade
GDP	gross domestic product
GFAR	Global Forum on Agricultural Research
GM	genetically modified
GMO	genetically modified organism
GNP	gross national product
GRP	Green Revolution Programme
HCH	hexachlorocyclohexane
HEI	higher education institute
IARC	International Agricultural Research Centre
IBPGR	International Board for Plant Genetic Resources
ICARDA	International Centre for Agricultural Research in the Dry Areas
ICLARM	International Centre for Living Aquatic Research Management (known as The World Fish Centre)
ICRAF	International Centre for Research in Agroforestry
ICRISAT	International Crops Research Institute for the Semi-Arid Tropics
IDRC	International Development Research Centre
IDS	Institute of Development Studies
IFAD	International Fund for Agricultural Development
IFPRI	International Food Policy Research Institute
IIMI	International Irrigation Management Institute
IITA	International Institute of Tropical Agriculture
ILCA	International Livestock Centre for Africa
ILRA	(merger of ILCA and ILRAD)
ILRAD	International Laboratory for Research on Animal Diseases
INIBAP	International Network for the Improvement of Banana and Plantain
IPGRI	International Plant Genetics Research Institute
IPR	intellectual property rights
IRRI	International Rice Research Institute
ISNAR	International Service for National Agricultural Research
LGA	local government area
MAFF	Ministry of Agriculture Fisheries and Food (currently DEFRA)
MANR	Ministry of Agricultural and Natural Resources
MRC	Medical Research Council
MNRR	Ministry of Natural Resources and Research
NACB	National Agricultural Cooperative Bank

NAFPP	National Accelerated Food Production Programme
NAP	Niger Agricultural Project
NAP	National Agricultural Programme
NARS	National Agricultural Research Station
NCRI	Nigerian Cocoa Research Institute
NERC	Natural Environment Research Council
NGO	non-governmental organization
ODA	Overseas Development Administration
ODM	Ministry of Overseas Development
OECD	Organisation for Economic Co-operation and Development
OFCOR	On-Farm Client-Oriented Research (part of ISNAR)
OFN	Operation Feed the Nation
OFR	On-farm Research
OPEC	Organization of the Petroleum Exporting Countries
OST	Office for Science and Technology
OVI	objectively verifiable indicators
PLA	participatory learning and action
PM&E	participatory monitoring and evaluation
PPA	participatory poverty assessment
PRA	participatory rural appraisal
PREST	Policy Research in Engineering, Science and Technology
R&D	research and development
RAE	Research Assessment Exercise
RAS	Royal Agricultural Society
RASE	Royal Agricultural Society of England
RBDA	River Basins Development Authority
RRA	rapid rural appraisal
RTD	research, technological development and demonstration
SAP	structural adjustment programme
SFI	sustainable financing initiative
SL	Sustainable Livelihoods
SPAAR	Special Programme for African Agricultural Research
SRC	Science Research Council
SSRC	Social Science Research Council
SRL	Sustainable Rural Livelihoods
SSM	Soft Systems Methodology
SSRC	Social Science Research Council
T&V	training and visit
TOT	transfer of technology
UGC	University Grants Council
UK	United Kingdom
UN	United Nations
UNDP	United Nations Development Programme
UNEP	United Nations Environment Programme

US	United States
USAID	United States Agency for International Development
WACRI	West African Cocoa Research Institute
WARDA	West African Rice Development Association
WTO	World Trade Organization
ZS	zonal supervisor

Introduction

This book came about initially as a result of informal discussions between the authors, all of whom are involved directly, to a greater or lesser extent, with agricultural science and research. With the public's perceptions in the United Kingdom (UK) fuelled strongly by media hype surrounding the issues of biotechnology, genetic modification (GM), bovine spongiform encephalitis (BSE), swine fever and foot and mouth it has become apparent that the level of debate in relation to agricultural research is becoming not only more and more polarized, but also increasingly politically charged and poorly informed. The arguments presented tend to involve the following themes. On one hand, it is put forward that farmers in the UK produce too much food, that it is environmentally damaging, that it is costly (in the case of the Common Agricultural Policy – CAP), that it is not healthy and that the whole system is supported by huge grants. Furthermore, it's said that this system of agribusiness is underpinned by research into new technologies that society does not want, carried out by white-coated boffins isolated in labs, who work for multinational corporations who collectively appear to have no social conscience or responsibility. The argument continues that this system of production is then exported wholesale to the developing countries of the world whether they want it or not. The opposing view is that food production needs to be increased to meet a growing population, that present methods of agricultural production are the only way forward to meet this demand and that new technologies being developed are not only safe, environmentally benign and equitable, but also transferable to the developing world.

It is no wonder then that discussions based around these two virtually diametrically opposed views have become increasingly heated. One of the consequences, so far, has been that both sides have tended to increase the amount of spin and PR in a desperate attempt to persuade society that it should decide who is right and who is accordingly, wrong, and thus apportion blame, right wrongs, and develop technologies that fit societies' needs and desires. However, is it really fair to blame farmers, agricultural scientists and large multinational corporations for the woes and inadequacies of agriculture and agricultural research and technology? It was with such thoughts in mind that it occurred to us that there are thousands of books, academic articles, journal papers, TV documentaries and press reports relating to all aspects of this topic, written and presented from every possible angle except that of the agricultural researchers themselves: who are they; why do they do agricultural research; what is agricultural research; what motivates them; who pays for it; and who decides what should be researched? It is these issues that form the backbone of this book. Furthermore, we realized that this was an opportunity to write about our work in an accessible way rather than presenting another purely academic text. To illustrate that we too need to voice our opinions in an effort to demonstrate that agricultural researchers, be they academics of the social or natural sciences or practitioners, all have a social responsibility for the work carried out, despite the pressures and contexts within which they work. In short, we felt it was time to put our case forward as objectively as possible.

THE APPROACH TAKEN

Agricultural research is an enormous subject area covering many different disciplines including: in the natural sciences – agronomy, soil science, plant physiology, hydrology, chemistry, biology, genetics, entomology, ecology and animal husbandry, to name but a few; in the social sciences – sociology, economics, gender studies, anthropology, history and geography (human and physical) and many more besides. Covering all of these topics in a book of this length would obviously be impossible. Therefore, the approach we have taken is illustrative and general rather than fully comprehensive. The overall aim is to inform and broaden debate surrounding agricultural research and what drives it.

The title *Science, Agriculture and Research: A Compromised Participation?* neatly sums up the approach taken. First, like all subject areas, agricultural research is prone to faddism. In other words, different methodological tools and approaches are used to identify, explain and investigate issues surrounding particular aspects of agriculture that are perceived to require attention, eg, increasing yield, enhancing environmental protection, securing the livelihoods of the rural population and so on. The underlying philosophies and social contexts of given times or places dictate which particular aspect is looked at and how it is investigated. Given that these parameters change, 'new' approaches come and go (and sometimes come back again, often under a new name!) as do the areas under investigation. Once the particular problem is identified (in a particular way) the next step is to instigate projects that address the situation. This book will explore how 'participation', the latest approach to be used in agricultural development in developing countries, has evolved. The underlying philosophy of participation is that it should be inclusive of the views, needs and desires of the supposed beneficiaries of research into agricultural systems, thus empowering people in the decision-making processes involved in project and technology development and implementation. The assumption is that, given this level of involvement the outcomes of the research will be more likely to fit the social context and therefore have a higher level of relevance for those it is intended to help. However, as will be illustrated, it is becoming apparent that this approach is fraught with complications and contradictions. Part of the explanation for this lies with agricultural scientists. Our hypothesis is that, given the pressures placed upon agricultural science, its practitioners and the underlying drivers or forces dictating the direction and funding of research, it is no wonder that 'full' participation is still proving to be an elusive goal to achieve. To understand why these pressures exist, the book will first examine what they are and who experiences them. Having set this scene, the book develops along two converging historical lines. One is the development of the farming system in the UK and its concomitant research base. The other is the exportation and development of this system to the developing countries of the world.

Chapter 1, 'What Drives Agricultural Research?', sets the scene by examining why agricultural research is carried out, who carries it out and what drives it. The emphasis is very much upon the present day situation in the UK. Chapter 2 examines the origins and development of agricultural research systems in order to further answer some of the questions posed by Chapter 1. In particular it reflects on the ways in which agriculture has developed over time and how research has driven and been driven by farming practices and contextual politics and social needs as they change over time. Chapter 3 explores the influences that the British Empire had in directing agricultural development and research in its colonies. The second half of the book, beginning with Chapter 4, concentrates on the rapid changes that occurred in agriculture beginning at the end of World War 2 and demonstrates how, with a supportive agricultural policy, farming in the UK became agribusiness. The chapter focuses on the growing institutionalization and politicization of research that occurred during this period ending up with the dominance of a system focused on biotechnological research that we have today. The irony of this would appear to be that, as we in the developed world have called for greater and greater participation of farmers in setting the research agendas in developing countries, we have at the same time been systematically eroding the participation of farmers and consumers in setting our own research agendas. Chapter 5 picks up the story of the development of agricultural research methodologies in developing countries from the end of World War 2. This chapter also further explores the apparent dichotomy presented in Chapter 4 by focusing on the drive to replace 'top-down' approaches in favour of the 'bottom-up' drive for participation. The validity of whether this approach actually does involve the resource-poor farmers and rural population as 'participants' in an agenda-setting process based on true equality is questioned. Finally, Chapter 6 ties together the themes developed in the book and raises some interesting questions in relation to the power relationships between those involved in agricultural research and development.

Chapter 1

What Drives Agricultural Research?

THE RESEARCH PROCESS AND POLICY – HOW DOES IT ALL WORK?

Having set the book in context in the Introduction, we would now like to explore the various factors that affect and influence agricultural research, from the political environment through to what motivates individuals carrying out the work. We can explore this with some very simple questions:

Why is agricultural research carried out? Is it to provide information to solve a problem, to develop new ideas or to expand on existing ones? Is it to 'move things forward', to provide opportunities for wealth creation, to 'keep ahead of the game'?

What drives agricultural research? Is it a need generated within the industry? And if so, at what level? Is the research self-motivating through the personal ambition and motivation of the researchers? Or is it politically motivated, either through national or international collaboration, or some technological need or support for industry? Or is it simply profit driven?

Who determines what drives agricultural research? Is it 'technology push' seeking to employ research through the development of techniques and materials looking for applications, or 'market pull' in which the need is identified by industry which then seeks a solution through research?

How closely is the research matched to the need? Who decides and how are the two brought together? Who monitors this decision and the provision of solutions through research?

Who ensures effective use of funds, time, labour and facilities, and who monitors this? Is there any accountability involved?

How are the objectives for the research set and who sets them? Who sets the objectives for the funding agencies and are the objectives based on requirements in the real world, perhaps in industry, or to serve a pressing need and, as such, recognized as being essential? Are the objectives influenced by the needs, aspirations or requirements of the scientist? Or are they politically driven, perhaps to satisfy a demand from the public, to gain parliamentary popularity, or for so called political correctness? They may even be strongly influenced by what is thought to be popular or trendy.

We will attempt to address some of these questions in this chapter, but first let us consider some contrasting scenarios to set the questions in context – an exercise which also helps to justify them.

Let's start with a popular perception of what research is about: the 'scientist-driven' agenda. This entails a scientist, wearing a white lab coat, slightly (or even very) eccentric, single minded and driven by an over-riding search for truth that keeps him working, usually alone, all hours of the day and night. Mostly, the nature of the problem is irrelevant – it is the image that counts and that has been used (and abused) all too frequently to advertise anything from mashed potato to cars, from toilet paper to cosmetics, providing an image of the high tech nature of the product, giving it respectability and making it appeal to the 'discerning' buyer who likes to 'keep abreast of the times' and who likes to be seen to respond to and embrace advances in technology. Matters of funding, career structure, teamwork, applicability and justification, and the list of questions posed earlier, do not enter into the situation at all – it's unreal, just for the media – and it can be very misleading if, as is often the case, this is the only portrayal of a scientific researcher some people ever see. Add to this the word 'agriculture' and we're usually in the world of talking animals, animated cows and pigs all striving to provide us with our daily food. Fortunately, agricultural research is rarely portrayed as such, other than through the production of food, so at least we are spared the vision of eccentric, white-lab-coated boffins with their wellies on!

But there is an important message here – that scientists are often perceived by the general public and especially by television audiences, as boffins, isolated from the real world and deeply engrossed in their all-consuming passion – research. By the same token, such research is usually seen as 'blue skies' or esoteric, detached from reality and carried out in isolation. For most research this could not be further from the truth.

So, let's get a bit more real. Consider another scenario with a research scientist working in a research environment, highly motivated and committed to a certain thesis, a particular idea he wants to check out within the sphere of interest in which he is working. He feels strongly committed to the idea that answering this particular question will provide valuable information and move the work along significantly. But he needs funding to carry out the work, so he looks around to see what is available and evolves a compromise to fit what he wants to do with the available funding, targeted at certain topics and research areas to satisfy the funding agency's requirements. Having obtained the funding, he may be free to pursue his own agenda, follow his own ideas and provide some innovative research, the outcome of which will be difficult to foresee, or he may be required to work to a strict plan with predetermined milestones and deliverables, based on agreed objectives set by the funding agency.

In contrast to this, consider a situation in which a crisis arises, such as a disease epidemic or catastrophic crop failure due to some phenomenon, or perhaps a long-term problem which has reached crisis point. The consequences of the problem may be far reaching, prompting government and/or relevant parties in industry to take responsibility for doing something about it. A strategy is formulated, funds are made available and a call for research proposals to provide information and solutions to the problem is put out inviting researchers to participate. In this situation the funding is provided for research which is strictly managed and carefully targeted to addressing the problem and the motivation is in finding a solution.

Another example can be envisaged in which a scientist proposes some good quality science targeted at a particular applied goal. The ideas are sound, the approach is both feasible and sensible, the costs are reasonable, and the project addresses the problem thoroughly with a high likelihood of a successful outcome. In reality, this is a very common situation and usually results in success all round. But the application for funding can fail at the evaluation stage due to competition from other 'high science' proposals which are considered by those in judgement to be 'better science'. This is all too familiar to researchers in the more strategic areas who are attempting to address particular questions in the real world. This is a topic we shall return to in the sections on funding and the funding agencies.

These examples illustrate the central and crucial dilemma in how research is funded. Should it be strictly managed, targeted to specific well-defined projects designed to answer specific predetermined problems in a highly controlled and managed way, or should it be left to the researchers themselves, as individuals or groups, specialized or multi-disciplinary, to determine their own targets.

In truth, of course, we end up with a mixture of these, resulting in varying degrees of control and freedom in what research is carried out and how, with most being somewhere in between the extremes of 'blue skies' and applied. But there remains the question of how it all works to the good of society and mankind and, in the case of agricultural research, for the good of the agricultural industry, from the breeders who provide improved varieties and strains right through the chain of producers and processors to the consumers and end-users.

We will explore this in a little more detail, but first let us compare the process, at least in the broadest terms, in developed and developing countries. In developing countries in the last 20 or 30 years there has been a strong tendency for research to be geared towards the needs of farmers – the farmer-first and participatory principles (to be discussed more fully in Chapter 5). Here, research is directed towards answering questions or solving problems encountered by farmers in a very applied, directed approach. In many ways, and perhaps ironically, this has come to contrast with the situation in the developed world where there is much more freedom, and indeed encouragement from various sources, to pursue more fundamental, innovative research – but perhaps this has lost its way in getting to the point where it often overlooks the needs of those in the agricultural industry, concentrating instead on, or being influenced by, more politically and publicly sensitive issues, determined by forces outside the industry itself. One of the fundamental differences between developing country and developed country research is that, in general, the latter is much more concerned with industry, since agriculture is industrialized on a large scale with few individuals managing large agribusinesses rather than by many individuals, each on a relatively small scale.

Take the situation in much of China or India, where 'farms' are usually no more than a few hectares with fields often the size of small plots, and where the agriculture is heavily dependent on labour input. Contrast this with some farms in Western Europe in excess of 500 hectares and individual fields in excess of 50 hectares – indeed, it is estimated that the minimum economically viable farm size in East Anglia in the UK, mostly down to arable crops, is now around 400–500 hectares. Such farms are often managed in groups of two or more and run very much as large businesses – and a far cry from the situation that prevailed as little as 50 years ago.

So the research requirements of the developed and developing countries are likely to be different, given this huge contrast in form and function – or are they? Let us return to exploring the research process. How do the policy-makers and funding agencies see research? Who do they think is in control? Who determines the objectives? Above all, and in relation to our concerns here, is the question of whether the process is responsive and, if so, to what or to whom, and if not, what does determine the objectives?

These statements relate to the questions of how policy-makers and funding agencies see research; what processes come into play in determining objectives and how they convince industry that it is worth investing in particular research areas. There may be some semblance of order in this, with procedures through which the process passes to arrive at the conclusion that this or that area should be targeted. There must also be some way to ensure fair play and good practice to avoid corruption.

The principles of how objectives for responsive research are determined are likely to be similar in both developed and developing countries, but the various factors influencing the process will be determined by the differences in the nature of how things are done and where and what the driving forces are. In the developed countries, these are likely to be industry and politics whereas in developing countries they are more likely to be farmer-driven.

Ideally, the process should be driven by information being made available on the nature of agricultural problems and what likely options are available to address them. To attract funding from industry or at least the attention of industry, industry must be convinced that the research is worth doing – it may not have to be immediately relevant or directly applicable, but it should have a high degree of relevance to its essentially commercial interests. Indeed, many researchers in industry can play a vital role in perceiving applications for more fundamental research. This is referred to later (Chapter 4) when considering the role of UK Foresight and exploiting the science base. But this is a 'ding-dong' process in which the different players strive to achieve their own particular objectives. It would be reassuring to think that they all had some common goal in mind other than furthering their own ends – something more idealistic and altruistic – but this seems unlikely, despite all the policy and mission statements about utilization and furthering the common good for example.

The policy-makers and funding agencies will have procedures in place, by which they determine where the funding goes and what areas require targeting. They will have access to information on what is required in practical terms, on the sensitivity of political and public attitudes and concerns, and they will attempt to ensure maximum benefit for the investment in research. But, of course, this may be flawed! It may be essentially naive or overly bureaucratic and full of misplaced motivations and commitment. And 'benefit' will almost certainly mean different things to different players in this game. The altruistic interpretation – for the benefit of mankind and the good of all – is almost certainly naive in practice, despite what many 'mission statements' would have us believe. It is far more likely that 'benefit' refers more to gain in some more worldly, materialistic and financial sense.

Similarly, consideration of what we mean by 'flawed' will also depend on perspectives, be they commercial, personal or academic. A decision which favours one party over another will be considered favourable to the beneficiaries but very likely considered flawed by the losers.

So the factors influencing agricultural research are many and varied, and certainly complex. The 'drivers' of research are difficult to define, determined as they are by the complexities.

THE IMPORTANCE OF FUNDING IN RESEARCH

One, if not the most important, determining factor in guiding research is funding. In this section, we will discuss sources of funding and how funding is determined, distributed and applied. Funding agencies usually identify amounts of their budgets which are to be made available for research (or research and development). These are sometimes under predetermined headings for specific areas or topics, and sometimes, though probably more rarely, made available for more speculative research. Research is usually expensive in terms of time, labour, equipment and facilities, and consumables.

As examples, we can consider two sources – industry-sponsored research and government-sponsored research. Industry-sponsored research can range from being very applied and directly relevant to industry's needs, to much more fundamental and speculative research (often called esoteric or 'blue skies'), which is not immediately relevant to the needs of industry but which explores areas thought to have potential, or is allied to current interests, or pursued simply through a fascination with 'how things work'. Industry-sponsored research can be carried out within the industry, using its own facilities and expertise, or, as is often the case currently, it can be carried out in specialized research institutes, either by providing funds as a grant for the work to be done, or on a strict contractor/customer basis. This decision may be on the basis of cost, quality of the work needed, access to appropriate facilities and expertise and the likely speed of progress for example. The work is usually highly targeted with clearly set objectives and time scales and is monitored throughout by frequent review meetings and progress reports; industry is understandably keen to get value for money and to see returns for its research and development (R&D) investments.

Government-sponsored research can be fundamental, applied or industry-driven, and much of the current government funding for agricultural research is often supplemented by input from industry in collaborative joint ventures (eg Link Programmes – see Chapter 4). Indeed, for some types of work, especially the more applied projects in which industry can be expected to have a major interest and play a major role, these collaborative projects are a most effective and successful way forward.

So where does the money come from to support expensive agricultural research? For government-sponsored research in the UK the main sources of funding are provided through the research councils (eg the Biotechnology and Biological Sciences Research Council (BBSRC – formerly the Agriculture and Food Research Council (AFRC), and before that the Agricultural Research Council (ARC)) and government departments, such as the Department of Trade and Industry (DTI) and the Department for Environment, Food and Rural Affairs (DEFRA – formerly the Ministry of Agriculture, Fisheries and Food (MAFF)); from industry it can be provided directly, on a customer/contractor basis or indirectly through government departments as in collaborative joint funding. A third and most important provider of funds for agricultural and related research, is that from private sources. These include philanthropic individuals, organizations or foundations, including non-governmental organizations (NGOs). Many of our now world famous research institutions were founded on money provided by such means and continue to receive at least some of their support from the foundations established at the time.

Most research these days is formalized, with procedures to be followed and applications to be submitted when applying for funds to enable the work to be carried out. But this was not always the case; indeed, formalized funding is a relatively recent phenomenon coinciding with the setting up of institutions specifically to carry out the research, such as those of the BBSRC. Prior to 1994, there were three times the number of research establishments, each specializing in particular areas of agricultural research and supported in the main by the umbrella research council. Most of these were established between the two world wars to cope with the needs of agriculture at that time and to exploit new techniques, methodology and technology, on the basis of both philanthropic inputs and government funds. But before this, agricultural research was usually supported by private individuals, who were often land owners seeking improvements in farming techniques and methodology (see Chapter 2).

Funding in some form or other is essential for research – it is the lifeline, for without it there can be no research. But how it is made available can have important consequences for the type of research, how, where and by whom it is carried out, and for the time scales and outcomes. Funding can be freely given, with no ties or expectations, thus allowing the research to follow its own path and the researchers to indulge in their own committed interests and speculations. But these days this is rare, although it is often the most common public perception of how research operates. This lack of accountability rests at the heart of the participatory movement – at least it did in its early days as it was thought that greater involvement of the supposed beneficiaries in the research process would yield better results. Funds can be provided as directed funding for specific work, topics or research areas, as is often the case with industry-sponsored funding and specifically targeted government funding, as well as the combination of the two in collaborative programmes. Directed funding can also be used to attract researchers towards specific areas of work, identified as being strategically important or directly and immediately relevant to the needs of the industry. This has increasingly been the way in which funding for the European Union (EU) framework programmes has been made available, by inviting research proposals addressing particular areas considered important by those formulating the programmes. How such target areas are arrived at will be discussed in the next section.

The different sources of funding influence the types of research and ways in which it is carried out. If the only funding available for agricultural research came from industry or was entirely industry-led confidential research, the benefits to the community at large would be greatly reduced since much of the information gained would be retained 'in confidence' by the companies. So although providing funds for applied, industry-directed research is obviously necessary in commercial terms, the freedom of alternative, non-confidential research provides an important balance of essentially 'free' materials and information, crucial for future developments based on chance discoveries.

WHO ARE THE FUNDING AGENCIES, HOW DO THEY WORK AND WHAT IS THEIR REMIT?

For most research and many researchers, funding agencies are the most important source of funds, providing the means by which the research gets done. This applies to both predetermined, targeted research carried out to strict plans and with strict objectives, through to more fundamental research with a higher degree of freedom. Funding agencies play a major role in guiding research, directing it towards specific goals either by placing closely monitored conditions on the release of the funds, or by defining areas in which investigation is considered necessary and inviting research proposals targeting these areas.

Given their important role in directing research, how do these agencies set their goals and objectives? How do they decide where to spend their money and in response to what? This could be in response to a real need identified as the result of observations and reports, and thus based on information received either from somewhere in the industry or amongst the research base; it could be as a result of issues and concerns raised through public concern and attitudes, such as the response to the BSE crisis and the introduction of GM crops; it could be through political pressures, again as in the case of BSE where the need for specifically targeted research became all too apparent as the crisis deepened. But it can also be in response to the need to provide a balance, to counter other sources that may be seen to swamp or over-influence certain areas. This can be immensely politically charged, affecting opinions and decisions, often on a very short-term basis akin to crisis management.

As research is so expensive and time consuming, most funding agencies devote considerable effort to identifying and defining areas requiring attention, although the basis on which the conclusions are arrived at will, of course, inevitably be biased by the interests of those individuals and companies contributing to the decision. Indeed, this very statement is often employed to stress the impossibility of truly 'objective' science. Methodologies may be objective in a sense, but decisions as to what they are used for are certainly not.

Two examples serve to illustrate this. The BBSRC boards and committees (see Chapter 4) are structured to ensure high quality scientific policy-making, planning and resource allocation. A single strategy board advises council on key strategic issues across the whole of BBSRC's areas of science. The board provides a high level overview for developing coherent funding strategies for research in universities and institutes. Working closely with and advising the strategy board are seven committees covering seven key areas of science: agrifood, animal sciences, biomolecular sciences, biochemistry and cell biology, engineering and biological sciences, genes and developmental biology, and plant and microbial sciences. These are intended to cover both basic and strategic science, and plan and propose new programmes both in the responsive mode and as coordinated programmes. The committees are responsible for peer review of research proposals, studentship allocations, programme planning, including identification of low priority areas in which funding may be reduced, and project evaluation. This structure provides the full integration needed to handle the continuum of scientific research from very basic and fundamental science through to that of a more strategic, applied nature. At the same time, flexibility is retained to support the best proposals in whatever area of science they might arise, and to develop coordinated programmes in areas of high priority to user communities, under a single focus in each of the seven key areas.

For a second example we can look at the EU framework programmes. Framework Programme 5 (FP5) is described as differing considerably from its predecessors, having been conceived to help solve problems and to respond to the major socio-economic challenges facing Europe. Its impact is maximized through focusing on a limited number of research areas combining technological, industrial, economic, social and cultural aspects. In this way, topics of great concern to the European Commission (EC) can be targeted by mobilizing a wide range of scientific and technological disciplines, both fundamental and applied, which are required to address specific problems, and which overcome barriers between disciplines, programmes and organizations. FP5 is directed mostly at strategic rather than near-market research (that is, basic research targeted towards applications rather than applied research per se) but there is an increasing emphasis on maximizing exploitation of research results and the transfer of technology into the market place. As a result, projects are expected to demonstrate a clear strategy for the subsequent development of their results into new marketable products, better manufacturing processes or other means of exploitation or dissemination. This is a clear indication that the EU sees the need to maximize the returns of its funding as part of its remit.

In addition to careful consideration of the objectives on which to spend their money, funding agencies also have to decide how to implement the process. This includes how the funding opportunities are advertised and how the applications for funds will be dealt with. This may be on a 'first come, first served' system, or involve a proposal process where applications are evaluated on the basis of the quality of the science undertaken and appropriateness in addressing the objectives, by panels drawn from the wider population of researchers with knowledge of the topic, or experts drawn from industry, academia and amongst policy-makers. In this way, the limited funds available are allocated to the 'best' proposals, the criteria for which determine what research gets funded. The application for funds can be in response to 'open calls' when research proposals can be sent for evaluation at any time, or in response to 'deadline calls' when applications must be submitted by a certain date. Various funding agencies use a mixture of these procedures depending on factors such as the urgency of the need to get the work in progress, and whether the interest is longer-term or ongoing.

The current trend puts far greater emphasis on commercialization of science and how the research feeds into industry (often referred to as 'technology transfer' or 'reduction to practice'), and what the benefits are to industry. These benefits are related to wealth creation (profit), whether directly through commercialization of products and new varieties for example, or indirectly by providing technologies or procedures that make commercial practices more profitable. This stress on commercialization for the objectives of research may come as a surprise to many with the commonly held belief that research is largely a 'blue skies' activity carried out by over-intelligent boffins in white coats. But increasingly over the years, and especially during the last 30 years since the Rothschild report (1971, see Chapter 4), agricultural research has become directed more towards commercial benefit (profit rather than discovery), and in doing so it has become increasingly reactive, having to respond to the perceived needs of industry and, indeed, politics. As research has moved towards providing solutions for industry, so industry has gained more control over the funding as a customer paying for a service, with attendant controls over what is done and how. Furthermore, as funding from government sources has become more concerned with wealth creation and value for money, it too has become increasingly responsive to pressures from industry and, more or less directly, political influence. Today's short-termism and the need for 'everything now' contributes to short-term thinking for both industry and government research, very much at the expense of longer-term strategic planning. Agriculture as an industry and its supporting research cannot respond to this short-term planning. With its long lead times it requires careful thought and long-term planning, supported by an acceptance that investments may not see returns for many years.

The funding is usually intended to cover research costs for fixed periods, often three years, but can be anything from a month or two to five years, but rarely more. Sometimes all costs are covered, including salaries of researchers appointed to carry out the work, and the costs of consumables and travel for example, but this is not always the case and funds may be limited to covering specific expenditure such as travel and subsistence. The different funding agencies deal with this in different ways, some always covering all research costs, others providing only partial support or start-up costs.

Thus, the funding agencies vary in their approach to supporting research, each dealing with the problem of what to support, for how long and to what extent, in their own way and with their own agendas for setting objectives and targeting specific topics.

WHO IS INTENDED TO BENEFIT FROM THE RESEARCH?

Having considered who benefits from the funding and how the process is applied, we now briefly consider who benefits from the research, noting that there may be a difference between who actually benefits and who is intended to benefit.

Those concerned with the distribution of funds will be anxious to ensure that the benefits of the research they are funding reach their intended targets in the broadest sense. Taking an example of research funding provided for crop improvement, the researchers will benefit from the funds in being able to carry out the work, to continue in productive employment and possibly further their careers. The information and materials this work produces may benefit a plant breeding organization in producing better varieties, which in turn benefits farmers and processors and the consumer with an improved end-product.

Thus, it can be seen that the beneficiaries from any particular funding can be many and varied, and perhaps beyond what might have been intended. We might consider how this has changed over the years, along the time-frame presented in Chapter 2. Has it ever been the case in the UK that agricultural research has been conducted on the basis of farmer participation and 'farmer-first' principles? The establishment of the research councils and associated institutes in the first half of the 20th century gives some cause to believe such a relationship did exist, but in today's climate of short-termism and responsive science greatly dominated by industrial and political influences, this must be rare.

So where does participation come into agricultural research in the UK, if at all? It seems doomed to be lost somewhere in big business and the quest for profit, which is contributing to agriculture's decline as an industry and the weakening of the farmer's voice and role, influence and persuasion.

Whilst permitting more 'blue skies' speculative research to be carried out, there has been a significant shift in emphasis towards 'wealth creation' as an important output and driver of research. This has one significant and potentially serious effect – that of losing much of the applied research to the confidentiality of industry. If industry is expected to support this applied work, it can also be expected to ensure it reaps maximum benefit which includes keeping ahead of rivals by maintaining secrecy. Thus, since it is likely that much applied research or even more fundamental work, funded and carried out under the control of industry, will have confidentiality conditions attached, the distribution and dissemination of results will be increasingly restricted, even to the point of delaying publication of the work in scientific journals for fear of rivals being able to gain from the investment.

So, who does the applied work now? Who ensures the transfer of technological advance into commerce? If left to industry, the requirement of confidentiality may greatly restrict the application of new technologies. But does this change depend on industry being better placed to fund its own research, or does it simply reflect changes in the way agriculture is carried out, being more industrial and like big business? This is, of course, often complicated by the varieties of industry that are involved in agriculture.

Returning to the beneficiaries of the research, there may be a contrast between who the funding agencies intend to benefit and who others (eg the scientists, public and research institutes) see as the beneficiaries. At least part of the answer to this is found in the reasons why scientists do what they do, and who they see as the beneficiaries. When asked why they do what they do, scientists are likely to respond with a range of replies, including: for the benefit of mankind (very worthy but probably rare!); as a job to provide the money to live; as a vocation, being highly motivated by curiosity; to satisfy the remit of departments and institutes; to benefit industry; to provide 'something useful' for society. Some of these may be somewhat removed from the motives and intentions of the funding agencies but, ultimately, it will depend on whether the work gets done or not.

THE PEER-REVIEW PROCESS AND RESEARCH FUNDING

Currently one of the most common methods of allocating limited funds is through peer review. This is a procedure in which research proposals and applications for funding are assessed by groups of experts drawn from the research community; that is, peers of the applicants and would-be participants in the proposed research. However, this can be very subjective and depend too much on the interests (vested or otherwise) and expertise of the individuals carrying out the assessments. A stark example of the peer-review process in action is the Eurovision Song Contest in which the extremes of voting and opinion stand testament to the problem!

However, despite its obvious failings, the peer review has become a generally accepted way of ensuring fair play amongst applicants, since it is intended to involve experts with experience and expertise relevant to the topics being proposed, and places the decision with these experts rather than with policy-makers who may be too politically or otherwise motivated. But there are drawbacks with this too, including the serious loss of confidentiality and novelty amongst competing peers, since those reviewing proposals have access – inevitably and by the nature of the process – to the ideas and information in the proposals. There is also the danger that the fields of interest and focus may become too narrow and specialized if decisions are left entirely to the experts and there may be the need for the assessment to be brought into a wider context. A further disadvantage, at least from the scientists' viewpoint, is that the process of assessment is time consuming and laborious, and takes them away from their research. This is particularly true where there are many proposals to be considered (eg EU framework programmes) or where particular, highly specialized expertise from a few individual scientists is regularly required.

Despite the drawbacks and the time-consuming nature of the process, most scientists favour the peer-review system, preferring to have, either directly by contributing to the assessment process or indirectly through having colleagues and peers on the review panels, some influence over the decisions. An alternative in which the decision is left to the politicians and policy-makers, with little or no scientific input, is certainly less attractive and, in practice, likely to be less motivating.

There are, of course, numerous other influences on the process of allocating funds, such as responding to personal, public and political opinion, with many examples in recent years in reaction to concerns over the BSE crisis and the introduction of GM crops. Another major factor is the need to match funds to the most suitable provider of the research; here however, other factors come into play, including political correctness (eg including certain countries as partners), satisfying public opinion (as in the case of redirecting funds to meet the needs of a crisis, such as BSE) and responding to crises, particular needs and requirements in the longer term, as with many environmental issues (such as the applications of bio- and phytoremediation in which microbes or plants capable of accumulating certain toxic chemicals are used to help clean up land contaminated with toxins such as heavy metals deposited as the result of mining activities or chemicals from industrial processing on so-called 'brown field sites'). With most funding agencies, there are two different processes at play, the identification of priorities, which is usually political and done by the funders themselves, and the selection of proposals, which involves the peer-review process.

THE FUNDING AGENCY PANELS AND BOARDS AND WHO INFLUENCES THEM?

It is worth pointing out at the outset of this section, that it is very much to the credit of the system that the whole process of funding is essentially transparent in countries like the UK. Panel and board memberships are published and readily available, and reports, evaluations and comments on proposals are fed back to applicants in helpful and constructive procedures. Negotiation, bartering, nepotism and corruption are, generally, not part of the funding process. Projects that receive funding are usually listed and available for public access; indeed, public reaction and adverse public comment are important concerns for funding agencies in making their decisions on what to fund.

The composition of the boards, panels and committees concerned with the allocation of funds is usually varied and includes a mixture of scientists (often eminent in areas of research relevant to the remit of the particular committee), representatives from industry (if appropriate) and the organization under which the funding is released (such as the research councils). Most funding agencies attempt a balance between practical knowledge and expertise, and eminence amongst panel members a, sometimes uneasy, combination of scientific and public respectability. This mix will often reflect the funding agency's particular remit or interest and will help to ensure that the funds are applied when and where the agency requires.

In addition to the input of these panels on the decisions about what areas and topics should be funded, whether in broad terms or more specifically, there is often strong influence from public opinion. It is very unlikely, for example, that areas of research considered by the public to be very unpopular or unethical will be listed to receive attention and some may even be withdrawn from the list in response to adverse public criticism. Indeed, there are a number of cases where committees considering areas for targeting have responded to these public pressures, recognizing and acknowledging environmental concerns, political motivations and, in some instances, media hype. Also, of course, it depends on the experts' interpretation of public opinion, which is not recorded as a matter of course and is often expressed only after the issues have arisen; much of the response to public opinion is based on the interpretation of a few individuals, perhaps as advisers but more often as members of the press and media. In recent years, public opinion in this form has become a major influence on what can and cannot be done, as in the great GM debate. However, most topics for research will not fall into these categories and will be little influenced by public opinion or involvement, and funding areas will be determined largely by the panels and committees.

Influences through industrial involvement will, again, depend largely on the topic, some will be great, others negligible. At least some of this influence will depend on whether scientists see applications and invite industry to be involved, accepting that this will bring with it constructive comments to influence what and how the research is done. The reverse can also be true where industry sees opportunities in academic research and invites input from the scientists.

An excellent example of how this operates in practice is the UK Technology Foresight Programme. The purpose of this programme is stated as being to:

- develop visions of the future, looking at possible future needs, opportunities and threats and deciding what should be done now to make sure that we are ready for these challenges;

- build bridges between business, science and government, bringing together the knowledge and expertise of many people across all areas and activities, in order to;
- increase national wealth and quality of life.

Supported by the UK government, it is intended to ensure that resources are used to best effect in support of wealth creation and improving the quality of life. Its purpose is to help business people, engineers and scientists become better informed about each other's efforts, and to bring these communities together through networks that will help to identify emerging opportunities in markets and technologies. The results will inform decisions on spending by government and industry, and the results are available to small and medium-sized enterprises that may not have the resources to undertake the surveys themselves.

As research in the developed countries has moved away from participatory involvement towards more political, industry-driven research in which seemingly ever-expanding projects address seemingly ever-increasing detail, the function of some of these panels is to maximize returns as wealth creation – which raises the questions posed earlier of who is intended to benefit from the funding and the research it supports, and accountability.

THE IMPORTANCE OF ACCOUNTABILITY AND VALUE FOR MONEY

Broadly we can recognize two contrasting types of research in this context. Fundamental research requires accountability in a managerial and strategic sense to ensure the most effective use of resources available, whilst applied research requires accountability in a 'money well spent' sense in meeting the objectives set to provide solutions to problems and in satisfying needs. In general, applied research with well-defined objectives and targets is easier to make accountable than basic research and is usually prompted by the need to answer particular well-defined questions. In contrast, esoteric/fundamental research, by its very nature of needing to respond to chance or unexpected findings, is less tangible in this sense and, therefore, more difficult to monitor and assess, although the accepted measures of accountability as effective and efficient working and managerial practices can and should still be applied. Industry on the other hand will ensure that research funds are closely monitored whether carried out 'in-house' or contracted out to research establishments. Indeed, industry will usually insist on close monitoring milestones and deliverables to keep the work on target, having set well-defined targets and time scales in the first place.

An example of procedures to ensure 'value for money' and maintaining accountability is to be found in the Research Assessment Exercise (RAE) which is applied to UK higher education institutions. This provides quality ratings for the research conducted in UK universities across all disciplines. It enables the higher education funding bodies in the UK to distribute, on the basis of quality and value for money, the public funds that will support future research.

It is deemed important that the users of university research – from industry, commerce and the wider community – are involved in the assessment of research quality under the RAE. This is intended to ensure the balanced assessment of research carried out in collaboration with, or directed towards the needs of, users, especially when the research is presented in a form that may be unfamiliar to some academic assessors (eg, some innovative products, designs or processes and new images or artefacts). The main purpose of the RAE is to enable the higher education funding bodies to distribute public funds for research selectively on the basis of quality. Institutions conducting the best research (judged, no doubt, in some tangible way) receive a larger proportion of the funds available to protect and develop the infrastructure for the top level of research in the UK. The RAE assesses the quality of research in universities and colleges in the UK by providing quality ratings for research across all disciplines, and takes place every four to five years. Panels use a standard scale to award a rating for each submission, according to how much of the work is judged to reach national or international levels of excellence. Outcomes are published to provide public information on the quality of research in universities and colleges throughout the UK. This information clearly has a much wider value than its immediate purpose since it can be helpful in guiding funding decisions in industry and commerce, charities and other organizations who sponsor research. It also gives an indication of the relative quality and standing of UK academic research. Furthermore, the RAE provides benchmarks that are used by institutions in developing and managing their research strategies. Across the UK as a whole, research quality, as measured by the RAE, is said to have improved dramatically over the last decade.

This is an example of one of the ways in which research quality, and strategic research in particular, is being assessed for quality and other benefits. There will, of course, be differences in the ways in which the assessments take place, and the methods and objectives used, and to whom the accountability is addressed. This is usually to the funder but it can also be either a representative or intermediary acting on behalf of the funding agency. Accountability may depend on the suitability of the work within research institute or departmental remits, even though the funding may be industrial or external. In this respect, it is most important that the research projects are suitably matched to the remit of the research establishment.

This becomes more important and increasingly difficult as the funding itself becomes more difficult to obtain and when scientists may be 'forced' to apply for funding for work not best suited to their circumstances, facilities or expertise. This can lead, for example, to research institutes that are better suited to esoteric research taking on increasingly applied projects, leading to a situation where important, more 'speculative', research that provides future opportunities is lost.

But, of course, there are problems with all of this. As with any assessment, its value will depend on the criteria used and, particularly when considered on a case-by-case basis, it provides a measure of 'performance' that undoubtedly greatly oversimplifies situations and circumstances. A seemingly trivial example in the grander scheme of things, but crucial to the career prospects of individual scientists, is the use of personal publication records (the writing and publishing of scientific papers) and the 'citation index' or 'impact factor' associated with different scientific journals. The number of publications is frequently used as the standard measure of output, and thus performance, of individual scientists and groups, and the quality of these publications is assessed on the basis of the 'quality' of the journals in which the papers appear, as judged by their citation index or impact factor. A list of journals has been drawn up with their 'value' indicated as the citation index or impact factor, relating to the frequency with which journals and papers are referred to by other authors. It turns out, not surprisingly, that trendy journals reporting cutting-edge, front-line science generally get high scores whilst those dealing with the less trendy (albeit important) areas get much lower scores. Scientists need to be aware of this – most are all too aware – and consequently need to choose to work in trendy areas such as biotechnology and molecular biology if they are to be supported. Research that supports agriculture more directly misses out here since this requires longer-term funding (see the next section) and usually involves areas of research, such as the more traditional approaches to crop and animal genetics, physiology and agronomy, not considered popular at present. This tends to result in researchers who choose to work in these areas being phased out in favour of more highly scoring 'high-tech' research. This is a real problem facing many scientists working in agricultural research at present.

THE MOVE TO SHORT-TERMISM

One of the most significant changes in funded research over the last 30 years or so has been the move towards short-term research – that is, research projects that require completion within a relatively short time span, usually three years or less. Whether this has coincided with, or been the cause or the result of, developments in biotechnology and related topics is difficult to ascertain, but the resultant combination has seen the demise of most longer-term research which, significantly, includes much that is associated with agriculture. Nearly all present day funding for agricultural research is directed towards projects of three years or less, encouraging work in areas with rapid throughput and results, and seriously discouraging work on topics such as crop genetics and physiology which, by their seasonal nature, are often restricted to one experiment (field trials or equivalent) per year. These areas have undoubtedly suffered as a result of most funding opportunities being directed towards short-term biotechnology projects. Of course, this is a sign of the times. Everything these days is short-term, disposable and instant but, unfortunately, agriculture is not like that and cannot respond to changes quickly – it requires planning, patience and forethought, all sadly in short supply in these modern times.

It is interesting to note how research councils and establishments and related funding agencies have accommodated these changes and how they justify the move away from supporting longer-term research to short-term research. This is referred to as 'underpinning science', a euphemism for avoiding commitment to long-term funding, and involves 'flexibility' in research programmes and strategies so that programmes can be changed at a whim. There can be no doubt that research related to agriculture and its associated activities has suffered as a result of these changes, with most research funding directed either towards short-period projects (of not more than three years) or to biotechnology, and mostly a combination of the two.

It will be interesting to see whether this situation is sustainable with the current objections to some of the applications of the new technologies to GM crops and farm animals since so much funding effort has been put into supporting these in recent years, especially by industry. Much of this has been industry-led, seeking what were thought to be lucrative GM products under the auspices of 'feeding a hungry world' but, at least for the present, the situation is on hold and the outcome is in the hands of public and political debate. There seems to be little hope however of returning to the longer-term funding of yesteryear to support agriculture, other than through these technologies so beloved of funding agencies.

The climate of short-term funding has other implications, too, most notably on the careers of most scientists forced to follow short-term contracts and having to move around to take up new contracts whenever and wherever they become available. All this, of course, is exacerbated by the demise of many 'permanent' positions in research, especially in the research institutes of the government research councils where these form a far smaller and decreasing proportion of the total workforce than 30 years ago. Thus, the career structure and opportunities for most research scientists are non-existent, which leads us back to the question posed at the start of this chapter of why scientists do what they do. These days it is unlikely to be for fame and fortune, but for most it remains a commitment to science (curiosity-driven or applied) and the desire to make a contribution. But for agricultural science this is becoming increasingly difficult. As with most things, there will be arguments for and against these changes, but many regret the losses to science that have resulted, especially to agricultural science.

Chapter 2

From Jethro Tull to Grain Mountains: The Origins and Development of Agricultural Research Systems

In order to further answer some of the questions posed in Chapter 1, it is important to reflect on the way in which agriculture has developed over time, and how research has driven and been driven by farming practices politics and the social needs of the time.

This chapter, therefore, represents a brief historical analysis and review of the changing pressures on agriculture and the research process, beginning in the 18th century, when some of the earliest records of a scientific insight into agricultural systems can be found. This history uses examples predominantly from the UK, though reference is made to incidents and developments in practice elsewhere in Europe and the US. Although there are differences in time and scale, the majority of the trends detailed here have been paralleled in other parts of the developed world.

A TIME-FRAME OF AGRICULTURAL DEVELOPMENTS SINCE 1700

1700   'Turnip' Townshend and the development of the 'scientific rotation' – the Norfolk four-course.
1701   Jethro Tull and the seed drill and horse-drawn hoe.
1762   The establishment of veterinary schools, firstly in Lyon, France, thereafter in Denmark, Germany, Austria and Hungary and in London in 1781.
        The establishment of the first board of agriculture for the dissemination of information pertaining to agriculture.
        The English Agricultural Society formed (became Royal Agricultural Society of England in 1940).
1841   The Irish potato famine.
        The founding of Rothamsted, the first agricultural research institute.

The establishment of the state veterinary service, following repeated outbreaks of rinderpest in cattle. The measures which were then introduced involving the slaughter of diseased animals. Severe restrictions on the movement and importation of animals led to the disappearance of the disease from the UK in 1877.

1866   Sheep pox eradicated from Britain.
The establishment of the Lawes Agricultural Trust – private funding for agricultural research.
The birth of plant virology – Iwanowsky discovered the filterability of plant viruses.

1897   Animal virology – Loeffler and Frosch filter the virus of foot and mouth disease.
Turn of the century: the age of the plant breeders – greatly improved varieties of the principal arable crops. Much more agricultural and biological research on the Continent than in the UK (for example, Pasteur, Koch and Ehrlich).

1902   The first commercial tractor produced in England.
1903   The National Fruit and Cider Institute at Long Ashton founded to provide region-specific research to the south-west.
Establishment of the Development Commission – its function was to advise the Treasury on making grants or loans for the development of rural areas, including grants for agricultural research.
The John Innes Institute founded.

1912–30 Eleven other institutes founded.
1914–18 World War 1.
1919   National Institute of Agricultural Botany founded in Cambridge to test new varieties.
1920s   The expansion of the synthetic ammonia industry for fertilizer production.
The establishment of the Agricultural Research Council for the coordination of agricultural research in the different research institutes.

1939   Outbreak of World War 2.
The establishment of the Agricultural Improvement Council as a mechanism for the transfer and application of research results.

1947–60 Post-war expansion and the setting up of specific units to deal with particular issues.

PROBLEMS WITH PRODUCTION

Agriculture has changed an enormous amount over the centuries and research has played a crucial part in increasing production levels. For example, average wheat production is now double that of the 1950s, three times that of the 1930s and more than ten times the level of the yields obtained before 1650 (Grigg, 1989).

Up until the end of the 17th century, agricultural practices in Europe had remained relatively unchanged for centuries, driven purely by demands for subsistence, the social hierarchy and changing demography. The development of an urban culture and the importance of cities as economic centres had perhaps had the greatest effect, but only in those areas close to the urban centres. In rural areas, practices were very deep-rooted and unchanging and there had been little or no increase in agricultural yields since medieval times. The average yield of wheat was fixed at around 10 bushels (262 kg) per acre (0.405 ha) and there were enormous seasonal and regional fluctuations. Farmers battled against the climate and weed and pest infestations, of which they had little or no understanding and absolutely no control.

John Wittewronge, the owner of Rothamsted Manor, during the 17th century, wrote in his diaries:

June 1684: A very dry season and excessive hot most of June. We had but 8 or 9 jaggs (ie cart loads) of hay this year out of Rothamsted mead [field].

June 1686: This was the greatest year of grass and hay that I can remember, for where I had the last year but 22 loads of hay out of all my meadows and uplands, this year I had at least 200 loads of hay, the last year I had but 8 loads of hay out of Rothamsted mead, this year 32 (Williams and Stevenson, 1999).

During this period, the UK was more or less self-sufficient in terms of agricultural production. The population was a fraction of the size that it is now and daily calorie intake per person was probably less than three-quarters of what it is today. In fact, for much of this time, England was a net exporter of cereals. Farms themselves were also more or less self-sufficient, with animals producing the fertilizer and power that was required and seed being saved from one year to the next. However, from 1740 onwards, the UK population began to rise sharply and by the end of the century, the situation had reversed, with about one-sixth of UK food consumption being imported. The rapid rate of population growth in the early 19th century meant that by 1841, one-fifth of food supplies came from abroad. By the end of the 19th century the UK was importing not just a great variety of tropical products such as tea, coffee, cocoa and fruits but also substantial amounts of cereals, dairy products and meat. Despite massive increases in production and the demands imposed by two world wars, Britain has never since been self-sufficient in basic agricultural commodities (Grigg, 1989).

Medieval agriculture was based very much on a system of open, or common fields that were owned by a single landowner and divided up into strips for cultivation by villagers. These strips covered all types of land from cultivatable land to wasteland and each villager would farm several strips scattered throughout the open fields. Up until 1600, the enclosure of these lands, by landowners, for specific purposes occurred gradually and in certain regions more than others. However, after this date the process occurred more rapidly. This was due to a number of social and economic factors, including the increased demand for grain from the growing cities, the demand for wool for the rapidly developing textile industry and its subsequent demand for industrial crops such as woad and saffron for use as dyes. Thus the products of the land became increasingly valuable to the landowners. The knock-on effect of this was that many villagers were turned off the land that they required for subsistence, to be replaced by sheep or highly competitive tenant farmers who farmed the land on behalf of the owners. This competition for land and production resulted in a major improvement in the efficiency of production, although this was at a very severe cost to the original and long-standing inhabitants of that land. A great deal of hardship and loss was inflicted on many people who were denied their traditional means of livelihood and subsistence. A number of peasant revolts were triggered by the effects of enclosure, with bands of rebels, known as levellers, attempting to claim back the land for the people. It was however, the breed of new tenant farmers who were later to become the devotees of a more scientific agriculture – as Gras (1940) has commented, 'victory belongs to the efficient cultivator regardless of his claim upon the soil'.

Over the years, agricultural enclosure has been portrayed as one of the great steps forward in terms of progress in agricultural efficiency and production. It is, however, interesting to reflect, in terms of the population as a whole and its well-being, whether it actually was the first step towards an agriculture that was inherently unsustainable.

THE AGRICULTURAL REVOLUTION: A THIRST FOR KNOWLEDGE AND ITS APPLICATION

The end of the 17th century saw the start of a new era of travel and exploration, and this brought with it new ideas and experimentation. Many gentlemen farmers had seen variations of existing agricultural practices being carried out in different parts of the world and they were keen to try them out. Most famous of these was Jethro Tull. His experience as a scholar at Oxford and as a landowner and his observations during a grand tour of Europe contributed to his invention (or rediscovery – he was probably not the first) of his famous seed drill. Perhaps even more important in terms of impact was Charles Townshend, who became known as 'Turnip Townshend', due to his pioneering work and his promotion of root crops in rotations. His rotation system was known as the Norfolk four-course, but also the 'scientific rotation' because it was based on an extensive study and experimentation that included crops unusual at the time in England and those brought in from the Continent, such as clover. These two and others like them, such as Arthur Young (1741–1820) and Robert Bakewell (1725–1795) championed new ideas and experimentation, stimulating a new interest in agriculture, its techniques and raw materials in influential and political circles and eventually bringing about some substantial and beneficial changes in practice.

This was also a time when lands overseas, outside of Europe, were being colonized and claimed by the Europeans. Much of the motivation behind this was the increased capacity to produce high value products through agriculture. Many governors of new states were landed gentry with an interest in the land and in production, so novel agricultural techniques and farming structures were rapidly transferred overseas to large-scale production of crops such as sugar, tea and rubber.

In 1836 Sir John Bennet Lawes, a later owner of the estate at Rothamsted, initiated some experiments on the manuring of agricultural crops. He had observed that it was common practice amongst the local tenant farmers to apply animal bones to the soil as a manure, but that their fertilizing effect varied considerably depending on the soil type. He initiated a series of experiments and set up a laboratory in his home at Rothamsted Manor in which he ascertained that by treating bones with acid, he could produce a much more powerful and fast-acting fertilizer than bones alone. In 1842, he was granted the patent for superphosphate and began a whole series of experiments aimed at exploring plant nutrition. He soon realized that in order to effectively address problems of agricultural production and to fully understand the complexities of plant nutrition he would need the collaboration of another scientist; he enlisted the help of Henry Gilbert, a young and enthusiastic chemist. Between them they tackled a vast range of agricultural subjects, making an enormous contribution to furthering understanding of plant nutrition and soil processes. They published papers on topics ranging from manuring practices to animal diseases and became recognized nationally and internationally for their work. It was one of the greatest and most influential scientific partnerships ever, that had one very strongly defining feature – that all of their work was targeted at the needs of farmers and the agricultural industry. Sir John Lawes became a very well-known society figure – so much so that a cartoon of him was published in *Punch*. He was also a great entrepreneur and scientist with very strong connections to agriculture and the farming landscape around him.

This was also the age of the great amateur naturalists and biologists and many great studies and theses were being initiated by wealthy gentlemen, clergymen and academics. Some of these studies included agricultural pests and diseases. In his book on farm insects, John Curtis (1860) described methods that had been tested to control various pests and noted that for *Calandra granaria* (grain weevil):

Fumigation with herbs having a strong and disagreeable odour – useless!
Turpentine – no good!
Sulphur fumes – inefficient because fumes did not penetrate grain!
Sifting the corn – no good!
Sudden heat killed but did not penetrate!

SCIENCE INTO PRACTICE – EARLY MODELS

The drivers of agricultural science at this time were very different to those of the present day, as discussed in Chapter 1. A common factor at this time was that the science was being driven by individuals who were educated and privileged enough to be able to devote their lives and considerable energies to subjects that intrigued or concerned them. They were frequently landowners themselves and possessed a strong social conscience and understanding of the agricultural problems of the day. Many were either politicians or were in a considerable position of influence and power. The importance of agriculture and its role in maintaining the economy and the livelihoods of the bulk of the population was clear to them – the majority of the population still worked on the land or were connected with farming in some way. Therefore, the communication of their findings to the agricultural communities was very important to them and many agricultural societies were formed for the purpose of disseminating this new information on techniques and practices. Below, in chronological order of founding are some of the national institutions with an express interest in agriculture and the dissemination of knowledge associated with it:

1660	The Royal Society
1754	The Society of Arts
1783	The Highland and Agricultural Society
1793	The Board of Agriculture
1799	The Smithfield Club
1804	The Royal Horticultural Society
1826	The Society for the Diffusion of Useful Knowledge
1831	The British Association for the Advancement of Science
1838	The Royal Agricultural Society of England

An interesting point to note here is that some of these societies, such as the Royal Society and the Society of Arts were (and still are in some cases) considered to be institutions associated with the scientific elite. Despite this, a recognition of the importance of application of science in agriculture is very apparent as demonstrated by the following examples. In 1664 the Royal Society established a Georgical Committee specifically for the study of 'agriculture and gardening', for which was produced a list of *Enquiries*, published in its *Philosophical Transactions*, dealing with best practice of agriculture in different parts of the country. These early agricultural surveys provided an enormous amount of information on practical aspects of agriculture. Even more strikingly, the Society of Arts provided grants for agricultural research, the first of which, announced in 1757, was for the best set of experiments on the nature and operation of manures, particularly emphasizing the practical utility of the science rather than just the theory as an end in itself. These examples serve to illustrate the elevated status of the practice of agriculture amongst the scientific and society elite of the day.

By the end of the 18th century, however, it was perceived that a public society, with government finance, would have more influence on agriculture than the existing privately funded societies and in 1793, the Board of Agriculture was established, with government support. Arthur Young, a prolific agricultural writer and commentator, who had promoted the cause of agricultural science within the Society of Arts and was a patron of the Smithfield Club, was appointed secretary to the board, despite some initial concerns as to its purpose and direction. These turned out to be well justified, because despite the best efforts of the board to engage the agricultural community through publications, the setting up of an experimental farm and the award of prizes for agricultural improvement, it remained remote from the agricultural community and was viewed with considerable scepticism. As an early government attempt for top-down control and focus of agricultural research and its communication, it was a complete failure and was dissolved in 1822.

The Royal Agricultural Society of England (RASE), which was founded as a private society, some 15 years later was established with some very clear objectives, which enabled the focus of activities and presented a strong message to both the farming and scientific communities that it was engaging: 'the acquisition of agricultural knowledge, the union of science with practice, and the communication of agricultural information' (Goddard and RASE, 1988).

Interestingly, the founders of the RASE, were convinced that 'an organization dedicated to the technical and scientific aspects of farming would only prosper if politics were rigorously excluded' (Goddard and RASE, 1988) and this policy was adopted by the society from the very beginning, despite, and perhaps because of, the fact the whole of agriculture was entering an immensely political and long-running debate over protection and support for farmers. This whole issue and its impact on farmers and innovation will be discussed later in this chapter.

The RASE meanwhile, quickly became established as the primary organ for the communication of technical advances to farmers in Britain, attracting more than 2000 paid-up members in less than two years. In the following years, membership peaked at around 7000, still a very small proportion of the total agricultural population and still dominated by the upper classes, but sufficient however, to ensure that messages and information were getting through. The major routes for dissemination quickly became established as agricultural shows and the publication of the *Journal of the Royal Agricultural Society of England*. Strong links with the farming and provincial press were established and maintained and were probably also very important in the general acceptance and recognition of the RASE as a source of useful information for all farmers. The RASE remains a very important society for the agricultural community and is responsible for major shows including the Royal Show, Cereals and Sprays and Sprayers.

SHIFTING POWER: THE DEMOCRATIZATION OF RESEARCH

Sir John Lawes, the founder of Rothamsted was very much in the mould of the great agricultural research pioneers of the 18th and 19th centuries. He was from a wealthy family of landed gentry. He was educated at Eton and Oxford and was driven by an intense scientific curiosity and desire to effect change. He financed his own research and that of his collaborator Sir Henry Gilbert through his own personal wealth and through the profits that he made on his patents and on the fertilizer business that he established as a result of his successful experimentation. He was a powerful and influential figure, well known in all the best academic, political and public circles. He was a very good example of the sort of character that had made such important strides during the years of the agricultural revolution.

By the end of the 19th century, power was beginning to shift from the aristocracy and was becoming firmly seated in the elected parliament and the government. With this, the ability of individuals to take forward their own ideas in science, became more limited and the implementers of research moved from the manors and stately homes to academic institutions and seats of learning. Research was still driven by individuals' curiosity, but additional pressures, drivers and funding were beginning to appear through the government and its associated agencies.

The Development Commission was established under the Development and Road Improvement Funds Act in 1909, with the role of advising the Treasury on making grants or loans for rural and coastal development. Its Development Fund rapidly became the largest single source of government support for agricultural research. The organization of such substantial funding under one government body changed the shape of agricultural research and focused it far more than it ever had been in the past. The Ministry of Agriculture was also influential in establishing specific laboratories in universities, and local government, in the form of county councils, also had a voice in the establishment of research facilities targeted to regional needs. The best example of this was the founding of the National Fruit and Cider Institute at Long Ashton in 1903.

Therefore, at this time, there was a shift away from individual endeavour in direct response to scientific curiosity and observed need, towards research driven by political ends. During the early part of the 20th century, this change coincided with a massive increase in demand for food from a burgeoning population and with the outbreak of World War 2 and the German U-boat campaign. Therefore, the political will was very strongly allied to a demand for increased food production, increasing yields and increasing profitability on farms.

At this stage, farmers and landowners were a very important part of the political scene. Agricultural organizations and farmers' bodies, such as the RASE, were held in extremely high regard and wielded a great degree of influence. With the Ministry of Agriculture, the Development Commission and the government as a whole all pushing for greater agricultural production in the UK, large amounts of public funding were channelled into agricultural research and between 1912 and 1930, no less than 11 agricultural research institutes were established.

This era, however, and many of the great scientific breakthroughs and discoveries that took place, initiated the process of the industrialization of agriculture and started a move away from a predominantly publicly-funded agricultural research base once again.

THE POLITICS OF AGRICULTURE – PROTECTION VERSUS IMPROVEMENT

The importance of agriculture in national security and the maintenance of the well-being of the population has always been a major driver in agricultural policy making, and in 1771, the first British import subsidies, known as the Corn Laws were imposed in order to provide incentives for farmers to produce for the national markets and to support them against competition from producers from other nations. This coincided with a period of increasing prosperity for farmers and of rising cereal and livestock prices in the UK. In fact, during the Napoleonic Wars, cereal prices reached remarkable heights in the general inflation and shortages of food supplies. In 1812, the price of wheat was nearly £2 per hundredweight, a price not reached again until 1953 (Grigg, 1989).

By the 1840s, with a burgeoning urban population gaining increasing political momentum, new ideas on agricultural improvement and concerns over the price of bread, the country became divided over whether the maintenance of the Corn Laws was beneficial to the population in general, or indeed to farmers. Politicians and agriculturalists argued over whether competition from exports would force farmers to adopt more efficient practices and take up new technologies, or whether it would lead to the break down of agricultural production in Britain and the dependence of the country on cheaper imports. In the end, the free-marketeers won through and the Corn Laws were repealed in 1846. It was in fact during this time of uncertainty that many important developments in agricultural technology (eg the development of artificial fertilizers) were initiated and the role of organizations such as the RASE for the communication of agricultural knowledge was recognized. In fact the 1850s and 1860s became known as a period of 'High Farming' in Britain and a time of agricultural development, innovation and prosperity, where even agricultural labourers experienced some gain in wages (Grigg, 1989).

However, by the 1870s prices had started to fall and cheap imports flooded into Britain. Farmers suffered considerably during this time, although British consumers benefited from cheaper food. By the start of World War 1 in 1914, the country was dependent on imports and made very vulnerable to attack on supplies, with resulting shortages and major hardship. The deep depression in the 1920s and 1930s led politicians, once again, to protect national farmers from competition. In 1924, farmers were exempted from local rates and in 1925 British sugar production was granted a subsidy. During the 1930s farmers were guaranteed a price for wheat and import quotas and tariffs were imposed on a wide variety of products including wheat and vegetables. The outbreak of World War 2 led to even further production support for farmers through subsidies and during this period, agriculture once again, prospered. Even more importantly, the political support for farmers was very greatly enhanced and avoidance of the dependence on imports during the early part of the century was seen as a major imperative.

In 1947, the Agriculture Act was introduced and between then and 1973, British farmers received considerable government support for production, through a wide variety of mechanisms ranging from guaranteed prices, to subsidies for liming, fertilizing and irrigation, as well as the imposition of import quotas and tariffs on most agricultural products. In parallel the newly emerging Common Market in Europe was developing similar strategies to protect European farmers and when Britain formally joined the European Economic Community (EEC) in 1973, farmers generally did not suffer financially in transferring from one system to the other.

Since before World War 2, British farmers have, therefore, been paid to produce, regardless of the market for their product. This has enabled agriculture to flourish in Britain and provided valuable support during the post-war years when increases in production were vital for the maintenance of the economy. However, this continued emphasis on production has had some major costs including the build up of grain and butter mountains that are only to be sold to developing countries at a later date.

By 1988 the EC was destroying 2.6 million tonnes of fruit and vegetables a year, comprising 23lb of lemons, 50lb of tomatoes, 41lb of oranges, 25lb of peaches, 24lb of apples, 15lb of mandarins, 7lb of cauliflower and 2lb of pears every second! (*Financial Times*, 1988). Furthermore, the EC stock levels for January 1991 stood at:

- beef stocks at 700,000 tonnes
- butter at 260,000 tonnes
- skimmed milk at 335,000 tonnes
- intervention stocks of cereals at 18 million tonnes.

These stocks were valued at ECU 3.7 billion in the 1991 European budget (Commission of the European Communities, 1991).

The system has greatly favoured larger farms over smaller units and increases in production have become the major goal for farmers as opposed to improvements in efficiency. This was in effect as a direct result of directing subsidies to production rather than any broader social or economic criteria. The agricultural environment has suffered as field sizes have increased, higher and higher levels of agrochemicals are used to push up yields and high-yielding varieties have totally dominated.

Production support has had a major impact on agricultural research in that the production increases demanded by farmers have been reflected in the research funding priorities of both government and levy boards. The real demands on agricultural systems have been obscured by the drive for production, such that the efficiency and sustainability of systems have been overlooked. In addition this may well have contributed to the increasing separation of the farmer and the researcher due to the fact that in the absence of competition, the link with technological advances becomes less important, so long as productivity can be maintained. This is in direct contrast with the period following the repeal of the Corn Laws during which time farmers recognized that in order to compete effectively they were going to need to maintain close contact with technological development. As is discussed in Chapter 4, agricultural research has been driven in recent decades far more by the availability of finance through industry or political imperative than by any sort of demand from the agricultural community. The current pressure, once again, for free trade imposed by the General Agreement on Tariffs and Trade (GATT), the World Trade Organization (WTO) and the enforced reforms of the Common Agricultural Policy (CAP), alongside the current debates on the impact of agriculture on the environment may be the factors that trigger a renewed emphasis on the relevance of agricultural research to farmers.

THE INDUSTRIALIZATION OF AGRICULTURE AND THE RESEARCH PROCESS

Goodman et al (1987) describe the process whereby the biological features and farm-based activities of agriculture have become converted into industrial processes, as 'appropriation'. These processes are then incorporated back into the whole agricultural production system. Probably the first and best-known instigator of this was Sir John Lawes, who very early on, recognized the potential of superphosphate production as an industrial process and a commercial venture. He took the practice of spreading bones on fields as a manure and through experimentation developed a process for producing superphosphate by treating bones with sulphuric acid. He took out a patent on this process and opened a works in Deptford to produce the fertilizer. By the 1850s, six superphosphate plants were operating in Britain and by the turn of the century, the fertiliser industry had become the major consumer of sulphuric acid. Similarly, the development of the Haber-Bosch process for synthetic nitrogen fixation and its subsequent scaling up to an industrial process in the early 20th century, established the heavy chemicals industry as an extremely important component of the agricultural industry as a whole and a major investor in agriculture-related research. Thus, advances in soil science had revealed an opportunity for industry that then itself became a driver for massive research investment.

However, many agricultural processes remained firmly in the domain of the biological sciences and research required public funding. Until the 1920s and the development of the hybridization technique on a commercial scale, the introduction of new breeds and varieties brought with it no commercial advantage to the developer, because seed from open-pollinated varieties could be saved and replanted. There was therefore, little incentive for private research. However, hybridization transformed this situation, principally due to the poor viability of double-crossed progeny, ensuring that new seed had to be purchased each year. Thus, by the mid-1950s the private sector had become the dominant source of new hybrid research (Goodman et al, 1987) and the research-oriented agro-biotechnology industry had been born.

CONTROL AND ACCOUNTABILITY

The early pioneers of agricultural research were accountable to no one but themselves. They did publish their research and were subject to criticism and review by their peers, much in the way that scientific publications are reviewed today. However, pressures on academics were very different and in particular the ever-present need to win and maintain funding from one body or other did not trouble them. With the growth of public funding for agricultural research, and the proliferation of agricultural researchers in institutes, a requirement to assess the quality of the research being carried out, and the need to ensure that it was directed towards the goals of the funders was recognized. However, it was not until the establishment of the ARC in 1931 that any coordinated effort was made to review or assess research progress in the light of the demands of the government or the industry.

Since then the research council responsible for the agricultural research institutes (the ARC, the AFRC, and subsequently the BBSRC) has held the responsibility for carrying out a comprehensive review of research at the institutes. Funding for the institutes has been dependent on the outcome of this review process and this has given the research council considerable power in directing agricultural research, even in instances where their contribution to the funding of specific areas of research has been in decline. The effect of the review process on researchers is further discussed in Chapter 4, but for the purposes of this historical reflection on the developing systems for agricultural research, it can be seen as yet another factor serving to distance the demands of the farmer from the pressures imposed on the scientist.

Chapter 3

Agriculture and the Empire: Transferring Technocracies

INTRODUCTION

The previous chapter has looked at the origins of agricultural research in the UK, and particularly how research became institutionalized in the early 20th century. There were a number of reasons for this, but one important factor was the desire of the government to ensure that the UK could be self-sufficient in food in an insecure world where war could break out at any time. But the UK also had an empire – the largest in terms of area, extent and population covered of any empire to exist on this planet. The 'official' period of the British Empire may have been relatively short-lived, but unofficial empire lasted much longer and has had major repercussions on the world we see today. Just as agricultural research became an issue in the British homeland, so it did in its colonies. Indeed the timing of both events is similar, yet the driving forces were in some ways quite different. The British Empire existed for commerce and trade, and the focus of the research was not on food staples but on cash crops that were luxuries. Sugar cane, tea, tobacco, oil palm and spices were cash crops that provided products not essential to life. The one exception to this was rubber – an important product in the time of war, but not indispensable. Although its empire has long since dissolved, the UK is now a member of the EU – the largest aid funder in the world.

Although the British Empire was relatively short-lived, it did have some major repercussions on the organization of 'formal' agricultural research (ie that carried out by paid researchers as distinct from what farmers do themselves). Indeed the institutionalization of agricultural research is an important factor in the broad sphere of rural development. It forms part of what some call the 'supply-side' considerations of research as distinct from the 'demand-side' (ie what potential beneficiaries want). This chapter briefly examines the origins of formal agricultural research in the British Empire, and the legacy it left at the time of independence of many countries in Africa (late 1950s to the 1960s). The aim is to illustrate the forces at play and how these may be seen as essentially 'top down' in nature with little involvement, if any, of resource-poor farmers.

This 'top down' emphasis existed for logical reasons. The research system was simply not created or operated for resource-poor farmers, and their participation was not an issue. However, and this is a key point, the reasons behind this are both logical and consistent within the wider environment of the empire and indeed the laissez-faire economic vision held by many throughout the world. As the sun set over the empire the repercussions of its agricultural outlooks and policies continued to have a strong influence, not least because many of the staff went on to work in the new government agricultural institutions and national/international agencies. In addition the physical and organizational structures left behind as the British withdrew formed the basis of the research institutions operated by the new states. Even as attention moved to food-crop research, in a local subsistence livelihood context, this was done within the inherited research structures and approaches that served the export-crop sector.

In this chapter we look at why, for better or worse, the developing countries inherited the agricultural research institutions they did. It is not our aim to pass judgement over the rights and wrongs of this evolution. Instead we explain what we think were the main forces at play and the repercussions these had for agricultural research systems in the newly independent states. Even with a focus on the British Empire, however, the problem with such a discussion is the danger of sweeping generalization over such complexity and variability. Nevertheless, such a focus allows us to explore the linkages between the evolution of agricultural research in a developed country, the British homeland, and some developing ones (the ex-empire, but now members of the Commonwealth). Not unexpectedly our thesis is that the two are closely related, but this is a point that is often missed or simply ignored. Even so, the heterogeneity of the empire does not easily facilitate the drawing of more general conclusions, so for that reason this chapter is more focused on Africa. However, even a focus on the British Empire can be dangerous given its heterogeneity, so in this chapter we have decided to first make some general points and then illustrate these through a single country case study: Nigeria. As well as highlighting some of the general points made, Nigeria also provides a number of exceptions to the rule that illustrates the danger of generalization.

Besides an analysis of developed–developing country relationships, there is a further reason behind our choice of the British Empire as a useful time-frame to explore. A growing focus on subsistence crops in the research sector of developing countries also coincided with a period of great change in biological research. In the 1950s and 1960s, the period when the empire in Africa died, the developed world in particular saw the birth of new outlooks and paradigms. Agriculture in North America and Europe became agribusiness as machinery, artificial fertilizer and pesticides were widely available and affordable. Ecology as a science reached maturity, and awareness of environmental issues began to impact on the minds of many and started to influence policy. These two events are not unrelated as problems resulting from poor soil conservation practices in the United States (US) and the widespread use of pesticides were major driving forces in this shift towards modern environmentalism. Lastly, but not least, was the birth of a new science that continues to have profound implications for the human race today: molecular biology and its newest offspring – biotechnology. Successive British governments have repeatedly stated their desire to ensure that the country is at the forefront of this new science and its application. One result of all this has been an acceleration of the already declining role for farmers in helping to direct research priorities, and a greater role for companies, politicians, pressure groups and scientists.

The chapter begins with a broad historical analysis of agricultural research in the British Empire; why it began and what it was for. We then move on to look at how this analysis relates to the experience of British Africa, but in particular to just one country: Nigeria. As part of this we look at the post-independence scene in Nigeria, including how it became part of an initiative to introduce a new Green Revolution into Africa. Included here are brief discussions of the evolution of agricultural extension and national and international agricultural research networks. Throughout this discussion we seek to illustrate some of the forces at play that may help to explain why the evolution proceeded as it did. The core relationship throughout is the one with farmers – the intended beneficiaries of the various interventions in agriculture. We also point to similarities and differences between this change and developments in the British homeland outlined in Chapter 2.

THE EMPIRE AND AGRICULTURAL RESEARCH

There is an immediately apparent difference between the UK and its empire with regard to agricultural research institutions. The latter were not established or shaped by the majority of farmers over a long period of time but imposed in an entirely 'top down' fashion by a foreign presence. One should remember that a major factor in the origin and development of the British Empire was not altruism towards others but trade predominantly in products from crops such as sugar cane, tea, spices, oil palm and tobacco. The phase often referred to as 'Informal' empire lasted up to the mid-19th century and comprised the extension of private trading companies throughout the world where they established local posts and an administrative apparatus. The navy and army presence in these far-flung outposts existed to support and maintain the companies' rights and privileges. Although the empire did not arise primarily out of a desire to 'do good' for fellow humans, evangelization did often go hand in hand with trade and no doubt some of those charged with governing the colonies did see themselves as responsible for 'helping' those they governed. Later apologists for the empire stressed its 'civilizing' role, but could not hide the fact that the empire was predominantly extractive in nature, and the slave trade of the Caribbean and America arose as a central element to this.

Generalization with something as large and diverse as the British Empire is dangerous. However, it can broadly be said that the commercial importance of export crops within the Empire naturally led to a desire to maximize production and quality. Crops such as sugar cane, cotton, tobacco, oil palm, rubber and tea were grown in plantations, and were owned mainly by European companies or individuals. Besides the obvious advantage that plantation crops tended to have a high value and hence repay investment in research, the plantation system had other advantages:

- Plantations are based on perhaps as few as one crop species, typically grown on the same piece of land, year in and year out (monoculture). It should be noted that one of the major disadvantages of crop rotation systems in general is that they may be considered to be 'wasteful of land' if part of the rotation period has to be given over to a usage which may not be directly profitable. Monoculture maximizes the return from a profitable crop and minimizes the need for a diverse range of machinery and inputs.
- The great uniformity of the crop allows almost total mechanization of land preparation, planting, maintenance, harvesting and storage. Indeed, uniformity is often a desirable characteristic during crop improvement programmes.
- A single crop species allows for straightforward management over large areas, and research on monocrops is much easier than looking at mixtures, in both time and space, of often quite diverse crop species.

It should be noted that initially food crops produced by the inhabitants of the countries within the empire were generally not regarded as important, other than the fact that they provided the subsistence for the workers who toiled on the plantations. This changed during the more formal period of empire in the latter half of the 19th and early 20th centuries, particularly in countries prone to regular famine conditions such as India and the East African colonies. However, for most of the period of empire there was no formal statement of government policy towards agricultural production in the colonies other than a fundamental and broad encouragement and protection of free-market trade. The first statement of government policy towards agriculture did not occur until 1945, and even then it was described as comprising 'a small publication... which was of little importance' (Masefield, 1972).

Staple foods for the British people were either grown in Britain, or imported from other European countries or, later in the 19th century, from the US. The colonies produced luxury goods not staples, although rubber later became a vital commodity in terms of war production (Masefield, 1972). In British West Africa, for example, this concentration on export cash crops, predominantly oil palm, as opposed to indigenous food supply was referred to as the 'Caribbean Model' (Sampson and Crowther, 1943).

A further important point to consider with regard to the early days of empire was the trans-continental movement of peoples. For example, African slaves brought with them the food crops they were used to growing, preparing and consuming. Supervisors and overseers, often slaves themselves, returning to their land of birth often brought with them crops from distant lands to be tried by their family and friends (Martin, 1993). The result we see today is the widespread cultivation of crops outside of their centres of origin. Cassava and maize are good examples, and the knowledge required for turning cassava tubers into a meal (gari) while at the same time removing the the cyanogenic glycosides was one of the most important innovations to occur throughout the African continent. Indeed, throughout the period of colonial rule almost all of the agricultural research meant to benefit resource-poor farmers was carried out on an informal basis by those same farmers. They carried on doing what they always had done, but with an increase of new ideas, mostly new crops, trees and varieties, brought about as an incidental (ie unplanned by the colonial power) effect of the widespread, and mostly involuntary, movement of their people. Not all of these introductions were positive. There were also accidental introductions of new and vigorous weeds as well as crop diseases and insect pests. Given that the rural communities, almost all of them farmers, made up the majority of the population in each colony, the arrival of European empires may have engendered some of the biggest changes in African agriculture seen on that continent for many centuries. But all of this was not planned or implemented by any colonial power for the greater good of the people it governed, but was driven by the very people they had conquered and enslaved!

On a more limited scale, the importance of export-crop agriculture within the British Empire inevitably resulted in a desire to look for improvements in production, quality, processing and transportation for example (Pardey et al, 1995). Just as in Britain itself, many of these early efforts were not part of a formal research process as we would know it today, but instead the efforts of individual plantation managers with commercial concerns very much in mind. Given that many colonial governors were also members of the landed gentry with a natural empathy for issues related to land use they also provided interest and support. The plough was tried in Jamaica in 1774 by a planter called Long, and Bennet wrote a letter in 1777 that describes his efforts at cotton seed selection in

Tobago. The latter described himself as 'being fond of experiments', and Masefield (1972) suggests that this may have been the first time that the word 'experiments' was used in connection with tropical agriculture. When the research process did become more formalized there was an understandable move to duplicate the sort of research processes developed to address issues in British agriculture. At first the emphasis was on the collection of new plants and taxonomy, and here Kew Gardens in London (established in 1759) which acted as a 'great exchange house of the empire, where possibilities of acclimatizing plants might be tested' (Masefield, 1972) was pivotal. Other botanical gardens along the lines of Kew were established throughout the empire to help with the study, testing and distribution of new plant species, particularly those that could have a commercial use. Garden staffs were also involved in some experimental work.

Like their British counterparts, specialized experimental stations devoted solely to agriculture in the tropics were an early 20th century creation. These stations tended to have a strong commodity focus, and a regional rather than national mandate (Pardey et al, 1995). Many of them were owned, managed and staffed by commercial companies rather than government, and some examples are shown in Table 3.1. Government stations did exist, but on a piecemeal basis. A report of the Lovat Committee, established to look at issues surrounding the organization of the colonial agricultural service and recruitment of officers, recommended in 1927 that a chain of agricultural research stations be established in the colonies. The Committee also recommended the establishment of an Advisory Council on Colonial Agricultural Research (ACCAR). The notion of a chain of research stations and an Advisory Council has many parallels with the later establishment in the 1960s of the largely US-funded Consultative Group for International Agricultural Research (CGIAR) and its global networks of stations. However the British government did not pursue the idea, and an ACCAR was not established until 1945.

Table 3.1 *Some Commodity-focused Research Stations Financed by the Plantation Industries in the British Empire*

Station and Commodity	Country
Tea Research Institute	Sri Lanka
Rubber Research Institute	Sri Lanka
Coconut Research Institute	Sri Lanka
Rubber Research Institute	Malaya
Sugar Experimental Station	Mauritius
Coffee Research Stations	Kenya
Sisal Research Station	Tanganyika

The research stations, and indeed the colonial agricultural service in general, were staffed mostly by European expatriates, with some 'educated' (in a British imperial sense) locals. The supply of 'suitably educated' staff for the agricultural service was problematic in the early years of the 20th century, partly because no university or college courses in agriculture existed in Britain. The Lovat Committee suggested the establishment of an Imperial College of Tropical Agriculture, and one was duly established in Trinidad in the early 1920s (now part of the University of the West Indies). The training included a period at Cambridge University where students learnt the latest techniques in agricultural experimentation and statistical analysis which were then very new subjects.

Colonial research agendas were heavily focused on production, and where important, quality (Guyer, 1993). Plant densities, maintenance procedures, breeding, use of inputs and irrigation for example have all tended to take centre stage when research priorities have been set, and those in charge of setting the agendas were the research station managers in conjunction with the plantation owners and managers. In contrast to the emphasis on cash crops for export, subsistence agriculture in the colonies received little if any attention within the research process. This is not to say that subsistence agriculture was not of interest to colonial agricultural officers or that plantation managers and colonial agricultural officers were not aware or ignorant of subsistence systems. The prevention and relief of famine was always seen as one of the top priorities of colonial agricultural officers, and they were required to tour their areas regularly and report back any signs of agricultural failure as a sort of 'early warning' system (Masefield, 1972). Hence there are many early examples of surveys and reports by such officials on more technical aspects of subsistence systems. By way of contrast, more sociological studies of households, for example, were carried out by the 'political' colonial departments rather than the 'agricultural' departments (Richards, 1983). However, although this reporting was important in terms of their day-to-day work, the officers were not generally proactive in terms of research geared at avoiding famine but reactive in that if there were signs of a problem then food could be redistributed. There were exceptions to the rule, and the Nigerian case study described below is one of them. On a more general level this relative neglect of food crops in research did change in the later years of empire after World War 2 when surveys suggested that malnutrition was more widespread than had been thought.

Besides the dominant desire to maximize production of exportable cash crops, a more marginal contributory factor for a lack of formal research on food crops may well have been the differences between agricultural systems in the colonies relative to the familiar (to the colonial agricultural officers) techniques of Europe. The crops, of course, were very different, but so were the ways in which they were grown. For example, bush–fallow rotations and intercropping (growing more than one crop on the same piece of land at the same time), the dominant systems of Africa, were alien techniques to the British. Yet these were also fundamental to survival in the African environment. Indeed to Paul Richards (1983): 'Intercropping, then, is one of the great glories of African science. It is to African agriculture as polyrhythmic drumming is to African music and carving to African art.'

The bush–fallow system would perhaps have been more familiar to European eyes as it is essentially a form of crop rotation with a fallow period (Figure 3.1). The use of fallow, mostly established grass/legume mixes, to restore soil fertility and reduce soil-borne pests and diseases was well known in Europe. In Africa, however, cultivated fields are simply allowed to revert to natural vegetation over a period of many years. The cropping period within the bush–fallow system typically involves an element of crop rotation that reflects a gradual decline in soil fertility. Therefore the African farm would be a patchwork of cropped areas and natural vegetation, with many crops mixed together in the same field. African farms did not, and in the majority of cases still don't, look like the defined fields and farm units to be seen in Britain. Plots are small and scattered, valuable trees are left standing in the plots and the crops often seem to be haphazardly thrown together in the same field. The overriding impression to those used to the agriculture of Britain may well be a 'mess', and the contrast with the organized and regimented plantation monocrops may well have been stark (Jones, 1936).

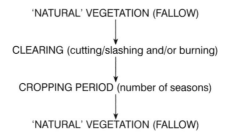

The cropping period may itself have defined sequences of crops and intercrops, and cultivation may be intensive:

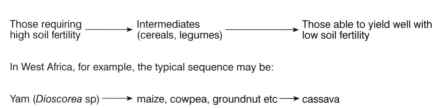

Figure 3.1 *Outline of the Bush–Fallow System of Agriculture*

As a result of this, African agricultural systems may not have appeared to be very productive to Europeans. The underlying rationale behind these systems was little understood, and an emphasis on low production as a perceived problem remained a central criticism of them from the perspective of colonial officials. Indeed, this perspective has been remarkably resilient despite a growing knowledge of the advantages that African agricultural systems possess. Even late in the 20th century some agriculturalists have suggested that agricultural systems are a representation of an evolutionary process, with European and North American systems, based on monoculture and high-input use, at the 'top' of the scale and African systems at the 'foot'. In the 'stage' approach to understanding agro-ecosystems, shifting cultivation was visualized as being at the low end of the evolutionary scale, with more intensive systems seen as the higher evolved form (Figure 3.2). Population pressure was often suggested as the main driving force behind this evolution. With such pressure, it is hypothesised, comes a need to move from extensive to intensive (Boserup, 1965). There have been numerous critiques of this vision, and some have pointed out that even in the bush–fallow system there may be an 'intensive' cropping period (Richards, 1983). Yet the dominant theme for much of the history of European involvement in Africa has been to 'improve' the African systems by making them more productive, and the usual way 'forward' has been to promote the methods of the Western world.

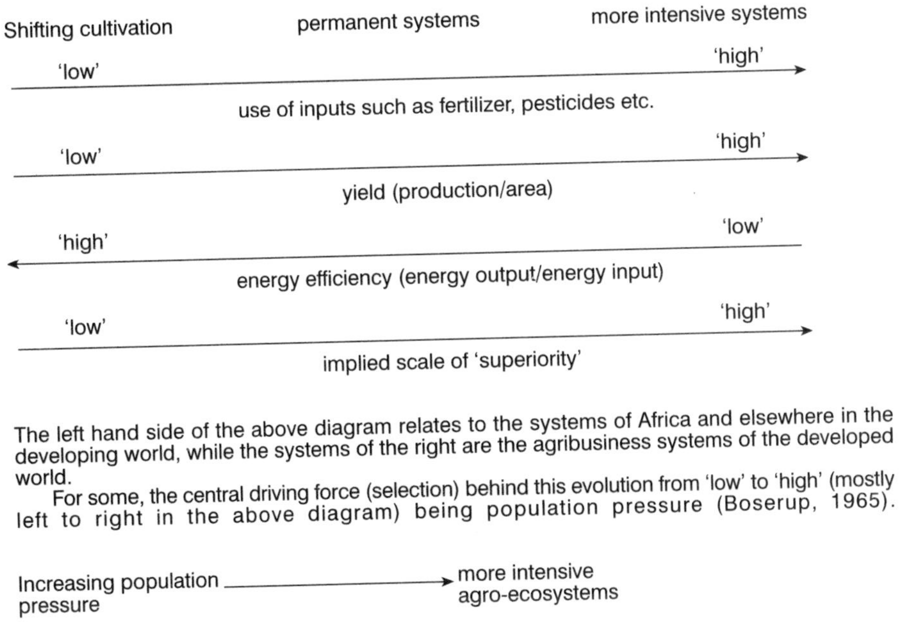

Shifting cultivation	permanent systems	more intensive systems

'low' → 'high'
use of inputs such as fertilizer, pesticides etc.

'low' → 'high'
yield (production/area)

'high' ← 'low'
energy efficiency (energy output/energy input)

'low' → 'high'
implied scale of 'superiority'

The left hand side of the above diagram relates to the systems of Africa and elsewhere in the developing world, while the systems of the right are the agribusiness systems of the developed world.
For some, the central driving force (selection) behind this evolution from 'low' to 'high' (mostly left to right in the above diagram) being population pressure (Boserup, 1965).

Increasing population pressure ——————→ more intensive agro-ecosystems

Figure 3.2 *The Stage Approach to Interpreting Agricultural Systems*

One manifestation of this is that as new agricultural technologies became available in Europe or North America they were immediately applied to solve production problems with cash crops in the empire. A good example can be seen following the advent of dichlorodiphenyltrichloroethane (DDT) in the 1940s. In 1945 a Colonial Insecticides Research Unit was established in East Africa to test and exploit the advantages that many saw cascading from the newly discovered organochlorine group of insecticides such as DDT. One can see the pace of change and enthusiasm for the potential of these new products from the proceedings of a conference held in 1953 (Wallace and Martin, 1954). The potential of DDT and related compounds such as gamma hexachlorocyclohexane (HCH) – also called benzene hexachloride (BHC) – for the control of hitherto uncontrollable pests of cotton, sugar cane, groundnuts and coffee are described. The low mammalian toxicity and good persistence of these insecticides in the tropics are presented as positive, and the excitement of having such new tools is tangible – even nearly half a century later. Beyond technical issues such as efficacy and dose rates, the problems discussed at the conference were basically those of availability to the peasant farmer of both the insecticides and means of application. The assumption was that making these products available at a cheap enough price would encourage their use and hence solve many of the perceived limitations to agricultural production. All this, of course, was before the serious environmental problems associated with these products became apparent throughout the 1950s.

Given this history, it was perhaps inevitable that when formal research in the British Empire did shift towards the more subsistence crops, the approach was much the same as it had been in the export-crop sector. Research focused on individual, or groups of related crops such as cereals, root crops or oilseeds, and decisions over what to work on were set by the European station staff. Unlike the stations that worked on cash crops and the agricultural stations in the UK the linkage between researcher and grower did not exist. One also has to remember that the productivity of agricultural research workers in the tropics was less than their counterparts in the UK research sector. Masefield (1972) gives a number of reasons for this, including:

- shorter working career;
- shorter working day;
- incidence of 'long' leave and sickness;
- fewer research facilities (because of cost, delays in delivery of material, unavailability of good assistance);
- lack of continuity in research (because of regular transfer, leave and sickness).

Perhaps the earliest example of a food crop emphasis was with rice in Malaya, but Masefield (1972) suggests that one of the best colonial programmes was the cassava breeding work of Storey in Africa. Given a production focus, the perceived constraint in Africa and the priority applied to cash crops and indeed agriculture in the UK, it is no surprise that studies of plant densities, input use (including insecticides), crop/animal nutrition and breeding, for example, all took centre stage.

Indeed, it should be remembered that in complete contrast to Britain the vast majority of agricultural research aimed at staple food production in the colonies was carried out informally by farmers and had nothing whatsoever to do with the formal research sector. The existence of this 'informal' research was often alluded to by those studying African agriculture, but was not fully appreciated until the end of the 20th century. Formal and informal research existed side-by-side with little or no interaction. This shouldn't be surprising given that the subsistence farmers had not asked for colonial government and neither had they been consulted about the form of colonial agricultural institutions and what they did. Empire was not a 'participatory' form of government! Given this history and relationship it is perhaps inevitable that even when their mandate broadened to include staple food crops much of what the research stations did was not necessarily relevant for the subsistence farmers.

Naturally with the increasing pace of independence from the late 1950s the colonial agricultural service declined in numbers and research stations were turned over to the new governments. However, the influence of the service remained. Staff transferred to other agencies involved in agricultural development such as the United Nations Food and Agriculture Organization (FAO), the UK Ministry of Overseas Development (ODM), which later became the Overseas Development Administration (ODA) and is now the Department for International Development (DFID), various Commonwealth agencies and the commercial sector. Some even returned to head UK-based and focused agricultural research stations. Perhaps most important of all in terms of influence were the many who took up contracted positions with the newly independent governments. In the early 1960s it is estimated that approximately 90 per cent of agricultural researchers in Africa were still expatriate, while in the early 1980s the corresponding figure was about 20 per cent.

For the most part the newly independent regimes maintained the status quo with regard to agricultural research structures, and in Africa the emphasis remained on cash crops (Roy, 1990). This is not particularly surprising given that:

- many expatriates remained after independence;
- close links to the colonial power were typically maintained;

- local staff were trained in the approaches and methods applied by the colonial power to research;
- there were many other pressing priorities besides agricultural research.

The last point is particularly relevant. Socio-economic and political turbulence has been the feature of many ex-UK colonies in Africa, and military regimes have been the norm rather than the exception. Besides insuring its own stability, government expenditure has tended to be focused on prestigious, highly visible and quick-fix infrastructure projects (bridges, roads, schools and hospitals, for example) rather than agriculture, although there has typically been no shortage of government exaltations with regard to the importance of agriculture. Rhetoric has been plentiful, while resources for research have been scarce. Indeed, investing public resources into agricultural research may have seemed like a luxury to many governments struggling to survive and industrialize as quickly as possible (Pardey et al, 1995; Nickel, 1997). Tribe (1997) estimated that even in the late 1990s most developing countries invested only 0.5 per cent of agricultural gross domestic product (GDP) in agricultural research:

Despite the demonstrated high payoffs to investment in agricultural research, and its vital role in rural income generation and conservation of natural resources, chronic underfunding threatens the survival of several research systems in Latin America and the Caribbean region. The situation in most African countries is as bad or even worse.

Such circumstances tend to favour the status quo, and further encourage a research infrastructure that mirrored that of the colonial regime that was created primarily to serve the interests of cash-crop producers and founded on the structures in the UK. The result was a very limited research infrastructure and negligible capacity to handle research on food crops (Yudelman, 1997). Yet research on subsistence crops was slotted into these structures. Allied to this was the typical attitude amongst the station-based researchers that African agricultural systems were low on the evolutionary scale, and hence little of value could be learnt from them. Understandably given this structural, historical and attitudinal environment the voiceless continued to be voiceless.

However, if more were needed, besides the issues described above, there were other pressures at play that may have hindered the inclusion of peasant farmers into the formal research process. Independence in many African countries occurred during the 1960s, and as already mentioned in Chapter 2 and developed more fully in Chapter 4, this period saw the rise of molecular biology and the environment as the new agricultural research frontiers in the Western world. In contrast, research into field-level agriculture in these countries became static before beginning a slow decline that continues to this day. These major shifts in outlook within the developed world were also highly influential in the developing world. While this is not, and should not, be irreconcilable with the notion of listening to farmers, there may be a tendency in the developed world to view African agricultural systems through these new mindsets. The result may be approaches looking for problems rather than the reverse: ie, 'How can we apply this new technology to solve …', rather than 'We need to solve …, how can we best do it?'. A desire not to be left out of these trends is understandable, especially given the potential power of some of the biotechnology, including GM, techniques. The dangers inherent in this have not been lost on some (see, for example, Meagher, 1990), although ironically the current shift away from GM in Europe and other parts of the world driven by the environmental lobby is now also being imposed on Africa. For example, we have the following quotation from Florence Wambugu (2000), one of Africa's leading plant geneticists, and based in Kenya:

> *Some aid workers here – I won't name them – are being pushed into an anti-GM position from their European office. They're being brain washed. We tell them we may not be the world's top scientists, and we know there are risks but we think we can manage them.*

CASE STUDY: NIGERIA

The principles highlighted in the preceding section will now be illustrated by more detailed reference to one ex-member of the British Empire in Africa – Nigeria. Although Nigeria illustrates many of the general points in the previous section, it also provides some marked and interesting contrasts as well as 'variation on a theme'. These will serve to reinforce some of the points made with specific examples, while at the same time highlight the dangers of generalization. For example, for reasons of climate and health West Africa never saw anything like the same intensity of European-managed plantations as did many parts of the Empire:

> *What distinguishes early Nigerian (and certain other West African nations) economic history from that of most other less developed countries is the fact that the foreigner arrived (and remained) not as a producer but as a trader* (Helleiner, 1966).

Therefore, Nigeria was regarded more as a 'protectorate' than a colony. Other than the above, the choice of Nigeria as a case study is not difficult to justify. The country has an estimated population of 112 million – about

19 per cent of the total sub-Saharan population in Africa – and hence the oft-quoted phrase that one in five Africans is a Nigerian. Indeed, Nigeria is one of the ten most populous countries in the world. It was one of the first African countries of interest to the UK – originally as a source of slaves, but later as a source of palm oil and other agricultural products – and was also one of the first countries to achieve independence in Africa (in 1960). Nigeria is also said to have the largest and most complex network of national agricultural research stations (NARS) in Africa (Hambly and Setshwaelo, 1997; Idachaba, 1998), with the possible exception of South Africa. In addition, Nigeria is the home of one of the agricultural research stations, the International Institute of Tropical Agriculture (IITA), that is part of a global structure (the CGIAR network). Indeed Nigeria was one of the first host countries to contribute financially to a CGIAR station within its borders. This allows us to discuss agricultural research within a national and international context. Nigeria also has a diverse agricultural landscape, with a variety of crops grown in a range of agro-climatic zones.

The initial interest of European powers (particularly Portugal, The Netherlands and France but later the UK) in Nigeria was as a source of slaves for the plantations of the Caribbean and the Americas. It is estimated that the trans-Atlantic slave trade accounted for the forced migration of perhaps 3.5 million people between the 1750s and the 1860s. Alongside this trade there rapidly grew a demand for palm products – primarily oil and kernels. The oil palm is an indigenous plant to West Africa, and the oil that can be extracted from its kernels has many uses. In Europe it was used to make soap and lubricants for machinery (before petroleum-based products were developed for that purpose). Because of the latter use, demand for palm oil grew during the industrial revolution in the UK, and ensured a strong British interest in this part of Africa even though the slave trade had long been extinguished. Yet the bulk of the palm oil was produced not by company-owned plantations but by the Nigerian farmers. The European-owned and managed plantation system never took hold in Nigeria, and even at the very end of empire it is estimated that less than five per cent of the cultivated area was farmed on a plantation basis (Oyenuga, 1967). The incentive for this local involvement seems largely to have been the hope of raising living standards, the acquisition of a wider range of consumer goods and perhaps an ability to invest in property. Hart (1982) even suggests that this 'standard liberal argument' is 'more plausible for West Africa than for almost any underdeveloped region'. Many changes in the structure and organization of local production and trade occurred as a result (Clarke, 1981).

Ironically, although the trans-Atlantic slave trade was banned in the early 19th century, the internal slave trade continued as a means of promoting oil palm and food crop production. The Niger river and the various streams at its delta became known as the 'oil rivers' as it was via these that much of the palm oil trading took place. As production was in the hands of the local farmers, food crops were grown alongside cash crops. Therefore unlike many other export crops in the Empire, from the very beginnings of the trade in Nigeria the export crop came from 'within the traditional world of the peasant' (Helleiner, 1966) and not some external and separate plantation-based system. Basically the palm oil trade was so well developed and efficient that European-owned plantations were simply not necessary (Williams, 1988; Martin, 1993). There were attempts by companies to acquire land for plantations, but these were firmly resisted by the colonial government (Francis, 1984). Indeed one company, Lever Brothers, had been persistently trying to acquire land for oil palm plantations between 1907 and the early 1920s (Williams, 1988). Besides the economic efficiency of local production there were other reasons for the unpopularity of plantations amongst the British. Fear of local insurrection was undoubtedly important, and the hazards of disease and climate in this part of the world certainly did not help (Burns, 1948).

Movement towards annexing this region as a formal part of the empire was slow, and really only occurred as a desire to protect trading rights from other European rivals, outlaw the export of slaves and maintain peace as a condition for trade. However, some did not see these advantages to the UK as strong enough to warrant the cost, and even after Lagos became a formal colony in 1861 a parliamentary report in 1865 urged withdrawal from West Africa (Burns, 1948). After all, the main British interest, trade, did not require formal annexation and, it was argued, could be left to the companies to enforce law and keep the peace at no cost to the taxpayer. In 1885 the Berlin Conference that allocated spheres of interest to the European powers stressed that only effective occupation would secure full international recognition, and hence secure trading rights. This reinforced the need for formal mechanisms, and the pace of change increased. The name, Nigeria, was adopted in the 1890s, and the different parts subject to British control were brought together into the Colony and Protectorate of Nigeria in 1914. Independence from the UK occurred in 1960, so Nigeria as an entity was only a part of formal empire for 60 years, although the Lagos colony was under formal control after 1861.

Oil palm was not the only crop of interest to the British in Nigeria. Timber, cocoa and rubber were also of major interest, as indeed were the annual crops groundnut and cotton. Trade in the latter two, which tended to be grown in the north of the country, was greatly enhanced with the opening of a rail link between Lagos and Kano (a city in the north of Nigeria) in 1912 (Helleiner, 1966). Animal diseases, particularly trypanosomiasis, that could be transferred to man also received much interest, and it was understandable given the rapid advances in medicine in the early 20th century that these diseases should receive much attention by a colonial power with economic interests in the region. Soon after the 1885

Berlin Conference Sir C A Moloney, the Governor of Lagos Colony initiated the Forestry and Agricultural Policy in the Colony of Lagos and its Protectorate (now the west of Nigeria) in 1887. In 1897 a Superintendent of Forestry was appointed, and in 1899, an Inspector of Forests was appointed by the Niger Coast Protectorate Administration (the east of Nigeria). The two Forestry Departments were united in 1906, and in 1910 the Department of Agriculture was carved out as a separate department. From the outset, the Department of Agriculture had responsibility for both cash- and food-crop production, although its activities largely revolved around the experimental cultivation of various indigenous and exotic crops and plants that may have had commercial potential.

Throughout the colonial period in Nigeria the centre of gravity of forestry and agricultural activity was firmly on the export crops and animal and human diseases. However, until World War 2 direct intervention in agriculture by the colonial authorities was very limited, and didn't go much further than work on experimental stations, provision of marketing facilities and inspection of produce prior to export. At the same time, it is perhaps ironical that increases in the export of agricultural produce from Nigeria were driven almost entirely by local producers responding to economic incentives. The British certainly helped this trade by the provision of infrastructure (storage facilities, rail and road links) and the introduction of new crops and varieties with export potential as part of its research programme, but local producers were responsible for increased use of both land and labour. In terms of cash crops, where production and quality were indeed paramount concerns, the strategies employed by farmers to achieve increases were largely their own. Given this it may be expected that the local producers would have had a bigger say in setting research agendas than elsewhere in the empire where commercial plantations dominated, but there is no evidence that this was the case. Farmers reacted to new initiatives imposed by the British rather than being involved in setting their form, but this was successful in the sense that production did increase.

Although the British involvement in agriculture in Nigeria was centred on cash crops for export, there is nevertheless an interesting – and unusual for the African colonies – parallel interest in local methods of food-crop production. The reliance on local farmers to produce the export crops, and hence the coexistence of cash- and food-crop cultivation on the same farms, may have been one catalyst for this interest. This was quite different from many other parts of the empire where export-crop production was entirely separate from food-crop production. Agricultural officials toured the colony, and many of their reports are still available. There were other interrelated factors at play. Richards (1985), for example, suggests that a major famine in the north of Nigeria may have been influential. Ironically this was partly related to the local emphasis on groundnut production for export even though some have argued that overall this commercialization of agriculture and the integration (regional, national and international) that came with it helped to reduce vulnerability of the rural population (Hart, 1982). In addition there is the influence of individual colonial officers; at that time a single individual could have a huge impact in policy terms. Between 1921 and 1936 Mr O T Faulkner was appointed as director of the Nigerian Department of Agriculture, and this marked an important change in policy. Faulkner was previously based in Malaya (1912–1914) and India (1914–1921). Unlike many other parts of the empire, Indian Departments of Agriculture had had to deal with a series of famines throughout the 19th century. This resulted in a greater emphasis on food crops for local consumption than in other parts of the empire, and colonial officers who served there were perhaps more attuned to the importance of food crops. Faulkner imported this emphasis into Nigeria, and saw the obvious juxtaposition of cash- and food-crop production in Nigeria as significant. Given that his first employment was in the research branch of the Rubber Growers Association in Malaya and that he has been described as a pioneer in field experiments, it is perhaps natural that he placed an emphasis on research station experimentation. Faulkner initiated a programme of station-based research that encompassed both cash and food crops, and upon leaving Nigeria in 1936 he was succeeded by his deputy (J R Mackie) who shared a similar outlook.

One of the main focuses for the Faulkner-inspired research programme in Nigeria was soil fertility and the bush–fallow system. As already described, it had hitherto been regarded as an 'inferior' system in terms of its perceived wastefulness of organic matter during bush burning, but there was a general realization that much more needed to be known about this system before suggestions for improvement could be made. In Faulkner's time the improvements were centred on the use of green manuring in the south of Nigeria and manures and mixed farming in the north. The idea was that once the local systems were understood and experiments conducted to test the efficacy of interventions, extension could be brought into play to spread the message. Faulkner's basic philosophy of understanding the local systems, experimentation and extension in that order may have been sound, but it was often forgotten by later generations of agriculturists in Nigeria. It is also noted that the research agenda was set by Faulkner and his colleagues. Western imperial eyes still saw the bush–fallow system as something that needed to be improved in terms of both production and soil conservation, and this outlook provided the foundation of the research programme. This work was spread over the 1920s and 1930s and accompanied by extensive extension and promotion, yet the technology was never adopted by the local farmers on a significant scale. Indeed, this is an interesting example as work with green cover crops has always been a favourite of researchers and development workers in Nigeria, and numerous trials have clearly shown a yield advantage as well as benefits to the soil. One form championed by IITA in the 1980s involved growing the crops between rows of leguminous bushes (alley cropping). To this day, despite all the research and promotional efforts stretching back to Faulkner, uptake of this technology in Nigeria has been limited.

Faith in the laissez-faire economic philosophy was greatly damaged by the Great Depression of the 1930s, and changed dramatically after the war. By the late 1940s and 1950s the colonial administration was planning and implementing grand and expensive schemes to increase production of export crops. The Niger Agricultural Project (NAP), focused on groundnut production, was one such scheme designed to meet a post-World War 2 need for food in Europe, and in particular a shortage of fats. This period saw an increase in popularity of centralized control of production based upon its perceived effectiveness during war time in Britain and western images of the Russian collective farms. Ironically in the post-war period measures to centralize production were more confined to the colonies than the British homeland. As described in Chapter 2, in Britain there were measures to widen the use of machinery and improve availability of extension, but the state never attempted to take control of the land. In the colonies it was a very different matter, and a classic example is the Gezira scheme for cotton production in what was then Anglo–Egyptian Sudan. The NAP was a later version of the Gezira scheme and was intended to follow a similar pattern, only with the focus being on groundnut for export instead of cotton. Ironically the NAP and its attendant philosophy were established in a country where a decentralized approach to export crop production had been highly successful.

The NAP was founded on the concept of transforming nearly 1050 square miles into something akin to the arable belts of the US and East Anglia in the UK. The project was planned in the late 1940s and clearing began in 1949, although only 10,000 acres were actually cleared during the lifetime of the project. Local people were to be housed in 'model villages', and encouraged to plant groundnuts as a cash crop for export as well as their own local food crops. Cultivation was to be mechanized and farmers were to be encouraged to use inputs such as fertilizer to maximize production. Even with the NAP the intention of the colonial authorities was not to establish a plantation, but instead the basis of the project was 'large-scale mechanized cultivation with peasant farmers occupying separate holdings but working cooperatively under supervision' (Baldwin, 1957). Although a reliance on local farmers for export-crop production had been successful in Nigeria, behind the NAP was the strong and widespread feeling that local African agriculture was unproductive and 'primitive' (Baldwin, 1957). In other words, Nigerian agriculture was at the low end of the spectrum in Figure 3.2.

Yet unlike Faulkner and his immediate colleagues, the NAP went further in not even appreciating a need to understand the local systems of agriculture – whether they were still perceived as 'primitive' or not. Ironically the 'modern' and 'western' vision of agriculture exemplified by the NAP was a complete failure. Tractors could not be maintained nor inputs supplied in enough quantities, yields were low and both soil erosion and land degradation were rampant. Indeed one of the main reasons given by Baldwin (1957) for the failure of the NAP was a lack of knowledge of local African agriculture:

> *The plain truth was that very little, if anything, was known before the scheme started about the existing agriculture in the Mokwa area, the reaction of new crops or varieties to local conditions, or the obstacles in the way of introducing techniques proved successful elsewhere.*

NAP did run a series of experiments on its own demonstration farm between 1950 and 1952, but these concentrated on the use of fertilizer and variety trials. They were not intended to improve knowledge of local practices and build upon them, but replace them. Indeed given the very nature of NAP it almost became '*one vast set of experiments*' that helped shed light on many of the problems that Nigerian farmers regularly faced (Baldwin, 1957)!

Nigeria has had a particularly volatile political history since independence in 1960, including a civil war between 1967 and 1970. This volatility has had an impact on the agricultural policy sector in general, particularly as other more high-profile projects tended to receive priority. Successive governments, often military, have had a penchant for creating, merging, cancelling and ignoring agencies and parastatals with an involvement in agriculture. Overlaps are created, often disruptive rather than positive, and it can be difficult to know just which agencies are operating in which part of the country. Some selected examples of the various agricultural agencies operating in Nigeria over a particularly volatile period since independence (the mid-1960s and the onset of Structural Adjustment in the mid-1980s) are shown in Table 3.2. Note that these are just examples, and don't include many other initiatives that impacted on agriculture such as the various Marketing Boards, Peoples' Bank and Land Use Decree (Francis, 1984). Throughout all of this, political concerns have been a major driving force for change, and there appears to have been little involvement of farmers in helping to formulate the various schemes from which they have been meant to benefit (Forrest, 1995).

Many of the government and internationally led interventions following independence continued to stress the importance of the cash-crop sector established during colonial times. Although food crops were included, they were more of a secondary concern with an emphasis on laissez-faire than intervention (Roy, 1990; Forrest, 1995). The discovery of oil in the country,

Table 3.2 *Summary of Agricultural Development Initiatives in Nigeria from Independence to Structural Adjustment*

Period	Initiative	Notes
1966	Ministry of Agriculture and Natural Resources (MANR)	
1968	Ministry of Natural Resources and Research (MNRR)	Agriculture deleted so as not to offend political sensibilities
1972	National Accelerated Food Production Programme (NAFPP)	Promotion of research, extension and services
1973	National Agricultural Cooperative Bank (NACB)	Funding of agriculture
1975	Green Revolution Programme	Modernization small-scale farming Mechanization (GRP) of small and large-scale farming. Spawned ADPs and RBDAs
1974 to 1985	Agricultural development projects (1st generation)	9 projects*
1978 to 1982	Agricultural development projects (2nd generation)	3 projects*
1983 to 1988	Agricultural development projects (3rd generation)	4 projects*
1976	Operation Feed the Nation (OFN)	Curbing of rural–urban shifts in population. Pride in farming and labour
1976	Nigerian RBDAs	Improvement of agriculture and seed multiplication in the major river basin areas of Nigeria. Control pollution
1984	Back to the Land Programme	Addressing the rising costs of food import bills
1985	Structural Adjustment Programmes (SAP)	Addressing the wider economic situation in Nigeria

* ADPs were all World Bank inspired and financed with a view to providing a solution to the food crisis in Nigeria

and its increasing production and export during the 1960s, 1970s and 1980s, meant that problems with food production could be hidden by imports. The 1970s saw something of a sea change with the adoption of a Green Revolution akin to that seen in Asia. Technological innovation generated by the formal research apparatus in Nigeria allied to an intensification of extension was at the heart of this (Roy, 1990).

AGRICULTURAL RESEARCH SYSTEMS IN NIGERIA

The Niger Agricultural Project (NAP) was in some ways a later manifestation of earlier initiatives to encourage improvements in local agriculture. The efforts of Faulkner and his colleagues have already been described, but from the early days of informal empire in the 19th century missionaries and other established model farms intended to demonstrate to the local people how to farm in the European mode. Indeed Nigeria saw possibly the earliest attempt from a colonial government at improving local agriculture which prophetically turned out to be a complete disaster. In 1841 a foray of four ships was sent to the town of Lokoja, which lies at the confluence of the rivers Niger and Benue. Lokoja was seen as the key to the exploration of the interior of Nigeria, and the establishment of a model farm was included as part of the mission (Burns, 1948). Land was purchased from the local chiefs for this purpose. However, after two months 48 of the 145 Europeans on the mission had died from fever, and one of the key personnel involved in establishing the farm had been murdered.

Gradually a network of NARS was established in the country, and perhaps because of the early start and emphasis it is said to be the most complex of such networks in Africa, with the possible exception of South Africa (Hambly and Setshwaelo, 1997; Idachaba, 1998). Once the North and South Protectorates were merged, agricultural research was centred at Moor Plantation, Ibadan (South West Nigeria) with branch departments at Samaru (Northern Nigeria) and Umudike (Eastern Nigeria). It should be remembered that agricultural research in export crops tended to be seen by the British as a transnational rather than national concern. Hence there was a West Africa Cocoa Research Institute (WACRI) with its headquarters in Ghana and a sub-station in Nigeria. After independence in 1960 the sub-station became the Nigerian Cocoa Research Institute (NCRI). In parallel with these stations was a trans-empire network of botanical gardens, with a part mandate for the testing and release of exotic material with potential for commercial exploitation. One important result of the emphasis on experimental station research in the early part of the 20th century was that Nigerian agricultural officers spent more time on research stations and less on local farms relative to their colleagues in East Africa (Forrest, 1981).

The Nigerian NARS network has undergone many changes up to the present day. A summary of some of the major stations is shown in Table 3.3, and the current research priorities set by the National Council for Agricultural Research are listed in Table 3.4. There is a clear emphasis on commodities rather than agricultural systems in this list, with the inclusion of the familiar export crops. Some have criticized the UK for lacking a clear vision of a Nigerian NARS, preferring instead to have a trans-empire vision encompassing all of British West Africa (Idachaba, 1998). Indeed this is said to be one ingredient in the apparent institutional instability seen in all of the Nigerian NARS in the latter part of the 20th century. It is certainly the case that all the stations in Nigeria were subjected to huge change both before and after independence, and a summary of these is provided by Idachaba (1998) in Table 3.5.

Institutions have been merged, split or their 'parent' ministry has changed. Indeed, it is perhaps more convenient to talk of the location of the research institute rather than its ever-changing name! The Agricultural

Table 3.3 *Origins and Locations of Some of the Nigerian National Agricultural Research Stations*

Periods	Stations	Location
1910–20	Development of Forestry (1912)	Ibadan
	Veterinary Research (1914)	Zaria
1920–30	Agricultural Research Station, Department of Agriculture (1922)	Samaru
	Provincial Experimental Farm (1923)	Umudike
	Food and Soil Research Unit of the Department of Agriculture (1924)	Moor Plantation
	Shika Stock Farm (1928)	Vom
1940–50	West African Institute for Trypanosomiasis Research (1947)	Kaduna
	West African Stored Products Research Unit (1948)	
1950–60	West African Institute for Oil Palm Research (1951)	Benin City
	Sub-station of West African Cocoa Research Institute (1953)	Ibadan

Research Station at Samaru, for example, merged with Ahmadu Bello University soon after independence in the early 1960s. At that time it was possibly the 'largest and most impressive regional research organization in the country' (Oyenuga, 1967). Yet surprisingly Samaru had little input into the work of projects such as the NAP in the late-1940s and 1950s, even though these projects partly failed through their poor use of fundamental research and local knowledge (Baldwin, 1957). In some cases the institutional change was designed to assist specialization. For example, Umidike now specialized in roots crops research.

Table 3.4 *Research Programme Priorities of the National Council for Agricultural Research, Nigeria*

Group	Crops / specific topics
Cereals	Sorghum, millet, wheat, rice, maize
Grain legumes	Soybean
Plantation/industrial crops	Cotton, rubber, coffee, tea, oil palm, coconut
Roots and tubers	
Livestock production	Veterinary, animal health / nutrition
Horticulture	
Fisheries	Inland fish, oceanography
Food technology	
Forestry	

Source: after Hambly and Setshwaelo, 1997

Table 3.5 *Changes Experienced by some Nigerian NARS*

NARS	Number of changes in parent ministry or supervising parastatal
Cocoa Research Institute of Nigeria (Sub-station of West African Cocoa Research Institute)	9
Nigerian Institute for Oil Palm Research (West African Institute for Oil Palm Research)	9
Nigerian Institute for Trypanosomiasis Research (West African Institute for Trypanosomiasis Research)	8
Institute for Agricultural Research (Agricultural Research Station, Department of Agriculture)	10
National Root Crops Research Institute (Provincial Experimental Farm)	11
Nigerian Stored Products Research Institute (West African Stored Products Research Unit)	9
National Cereals Research Institute (Food and Soil Research Unit of the Department of Agriculture)	10
National Animal Production Research Institute (Shika Stock Farm)	11

Note: Previous names given in parentheses
Source: Indachaba, 1998

NARS have also not been immune from all the overlap and change in the broad agricultural sector outlined earlier, nor indeed have they been any better than these disparate groups at including end-users of the research in setting agendas. Idachaba (1998) suggests that end-users have contributed to instability:

> *Users and beneficiaries of agricultural research who have a lot to gain from institutional stability have failed to articulate the demand for stable institutional arrangements from the government.*

While this is undoubtedly true, as we have seen it is not a new phenomenon in Nigeria and in the historical and institutional context is perhaps understandable. Indeed, one could also argue that the NARS and all the other parastatals and agencies themselves have not helped foster such a relationship, but again in the circumstances this is perhaps understandable. After all, seen from the point of view of these agencies it may be suggested that surges in funding or lack of it, job insecurity and changing mandates that come with different parental ministries, for example, do not help foster a more long-term perspective that includes having farmers as partners.

Yet Nigeria has one major advantage over many other African countries: it has within its borders the International Institute of Tropical Agriculture, part of the CGIAR network already mentioned, with a history dating back to 1967. In theory at least this institute should be immune to much of the instability that has gripped the national agricultural sector. Funding comes from outside the country, and salaries are higher and more stable than those of the NARS. Is there any evidence to suggest that this institute has been any better at including farmers and other stakeholders in setting its research agenda? This is a point that will be returned to later in this chapter.

AGRICULTURAL EXTENSION

As well as the research base, some mention should also be made of extension activities in Nigeria. The importance of extension as a means of communicating the results of research has a long history, but perhaps the first use of the phrase 'extension work' in the British literature of tropical agriculture can be found in an annual report of the Nigerian Department of Agriculture for 1928. This may be unsurprising given the strong views of Faulkner on the importance of adequate research prior to intervention. In later years extension systems in developing countries, including Nigeria, became more formal and structured. Probably the classic example of this is the Training and Visit system (T&V) of extension. T&V is an example of the transfer of technology (TOT) model of extension put into practice. TOT is essentially the top down transfer of a 'valued' knowledge (from the perception of those that have developed it) to recipients, or customers, who are thought to need it (Garforth, 1997). There may be slow adopters, but TOT assumes that new technology developed by researchers will eventually diffuse through the customer body. TOT has long been the dominant approach in extension.

The T&V system had its origins in the Middle East and was applied on a large scale in Asia (Garforth, 1997). Perceived successes lead to it being championed by the World Bank, and primarily for that reason the T&V system was introduced to Nigeria on a large scale during the wave of World Bank Agricultural Development Projects (ADP) in the mid- to late-1970s. Figure 3.3 is taken from one of the first generation projects; the Ayangba Agricultural Development Project (AADP) that existed between 1977 and 1982 in an area adjacent to the confluence of the rivers Niger and Benue and close to Lokoja. Divisions and zones are geographical entities typically based on local government areas (LGAs) and accepted divisions within LGAs based on chieftaincy. The World Bank projects in Nigeria were large-scale extension projects allied to supply (seeds, fertilizer and pesticide) and infrastructure development (roads, boreholes) (Forrest, 1981).

The early generation of ADPs had a very limited mandate for research, and was meant more as a vehicle to carry the results of research from the NARS and elsewhere to the farmers. Participation in this context was

Figure 3.3 *An Example of a Transfer of Technology Model of Extension Based on the T&V System*

simply measured in terms of adoption rates by farmers of the improved agricultural technology (Forrest, 1981). It is also perhaps ironical that one of the main activities of the ADPs was centred on the creation of land development schemes in exactly the same mode as the colonial NAP established in the late-1940s. Predictably given the NAP experience the results were exactly the same.

Although the T&V system was claimed to have advantages (being intensive, with a clear management structure, and accountability) it was also recognized that any extension system needs to maintain an upward flow of information from farmers to researchers.

> To remain effective, extension must be linked to a vigorous research program, well-tuned to the needs of the farmers. Without a network of field trials upon which new recommendations can be based and without continuous feedback to research from the fields, the extension service will soon have nothing to offer farmers, and the research institutions will lose touch with the real problems farmers face Benor and Harrison (1977).

In practice, however, the very hierarchical structure of T&V does tend to place farmers at the bottom of the hierarchy and researchers at the top!

Although feedback from farmers to the top of the hierarchy and beyond is encouraged in theory, in practice it tends to be minimal and the emphasis instead is upon a one-way flow of information and recommendations. The result is that farmers are seen as passive recipients of valued knowledge while they have little knowledge of use to researchers. Extension in the ADPs had much the same mindset, and the prime focus was on increasing agricultural production by some notional target during the lifetime of the projects. World Bank targets at that time were almost entirely based on an increase in agricultural production above a notional baseline figure, such as 15 per cent. Indeed, the Nigerian ADPs can be seen as a local manifestation of a more general drive during the 1970s to transfer the results of the Green Revolution in Asia to Africa. Amongst the major players in the origins of the Green Revolution are members of the CGIAR system, and it is perhaps not irrelevant in terms of the origins of the ADPs that Nigeria is host to one of the CGIAR stations.

THE CGIAR AND AFRICA'S GREEN REVOLUTION

The CGIAR held its first meeting on 19 May 1971 at the World Bank, Washington, DC. Eighteen governments and organizations attended the meeting as members, but none of them were from a developing country although many joined in later years. A chronology of membership is shown in Table 3.6. In 1975 Nigeria was the first developing country to join the CGIAR.

The prime thrust of the CGIAR was to increase food production, with the initial priority on cereals but eventually covering 27 commodities. The intention throughout was to work with NARs and at the same time strengthen their research capabilities. However, it was perceived that throughout the developing world there was a shortfall in nationally funded agricultural research capacity, so one of the prime outcomes from the early meetings was to create a world-wide network of internationally funded research stations, the International Agricultural Research Centres (IARC), to fill this gap (Tables 3.7 and 3.8). Although, like the early imperial station networks described above (eg the ACCAR), each station had a crop and regional focus, the founding resolution of the CGIAR declared that 'account will be taken not only of technical, but also of ecological, economic and social factors.' As a result the CGIAR branched into other areas of activity such as livestock, policy, farming systems and water management, and the number of stations increased from the initial four to its current sixteen.

Given the high profile enjoyed by the CGIAR and its IARCs it is sometimes easy to over-estimate their influence on a global scale. Yudelman (1997), for example, estimates that in the mid- to late-1990s only 4 per

Table 3.6 *Chronology of membership of the CGIAR*

Year	Members
1971	Belgium, Canada, Denmark, France, Germany, The Netherlands, Norway, Sweden, Switzerland, UK, US, Asian Development Bank, Inter-American Development Bank, International Development Research Centre (IDRC), United Nations Development Programme (UNDP), World Bank, Ford Foundation, WK Kellogg Foundation and the Rockefeller Foundation
1972	Australia and Japan
1974	United Nations Environment Programme (UNEP)
1975	Italy, Nigeria and Saudia Arabia
1977	Arab Fund for Economic and Social Development and Commission of the European Communities
1978	African Development Bank
1979	Ireland and the International Fund for Agricultural Development (IFAD)
1980	Mexico, Philippines and the Organization of the Petroleum Exporting Countries (OPEC) Fund for International Development
1981	India and Spain
1984	Brazil, China and Finland
1985	Austria
1991	Luxembourg and Korea
1993	Indonesia
1994	Russian Federation and Colombia
1995	Bangaldesh, Egypt, Iran, Kenya, Romania and Syria
1996	Ivory Coast
1997	Pakistan, Republic of South Africa, Portugal, Peru and Thailand
1998	Uganda

cent of the global agricultural research capacity resided in the IARCs. The developed and developing worlds were each estimated to have approximately 48 per cent of the capacity. Given that the IARCs were intended to fill a gap in nationally funded research, there have been recent calls for funding from international donors to be redirected to the NARS (Yudelman, 1997) although others see this as undesirable (Nickel, 1997). Indeed, the core budget of the CGIAR network has seen a gradual and sustained decline although the reasons for this are perhaps more complex (Hartmans, 1997). Nickel (1997) describes the problem succinctly and brutally, although in a form that is somewhat simplified:

> *It has become too large, too bureaucratic, and too politicized. The excitement and youthful agility it possessed when it began in 1971 have vanished; it has become a 25 year old that is arthritic and verging on dull senility.*

One of the IARCs, the International Institute for Tropical Agriculture (IITA), is based in Ibadan, Nigeria, within easy reach of the first capital city, Lagos, and geographically close to some of the first agricultural research

Table 3.7 *Annual Budgets (1993 Data) of the CGIAR Research Centres*

Name	Acronym	Core budget (US$ million, 1993)
International Rice Research Institute	IRRI	25.8
Centro Internacional de Majoramiento de Maz y Trigo	CIMMYT	24.1
International Institute of Tropical Agriculture	IITA	21.9
Centro Internacional de Agricultura Tropical	CIAT	25.5
International Crops Research Institute for the Semi-Arid Tropics	ICRISAT	26.9
Centro Internacional de la Papa	CIP	15.1
International Laboratory for Research on Animal Diseases	ILRAD	10.9
International Livestock Centre for Africa	ILCA	13.5
International Board for Plant Genetic Resources	IBPGR	9.0
West African Rice Development Association	WARDA	5.2
International Centre for Agricultural Research in the Dry Areas	ICARDA	16.2
International Service for National Agricultural Research	ISNAR	6.6
International Food Policy Research Institute	IFPRI	8.3
International Centre for Research in Agroforestry	ICRAF	11.9
International Irrigation Management Institute	IIMI	6.8
International Centre for Living Aquatic Research Management (known as The World Fish Centre)	ICLARM	4.2
International Network for the Improvement of Banana and Plantain	INIBAP	2.1
Centre for International Forestry Research	CIFOR	3.4

stations established by the British in Nigeria such as Moor Plantation. IITA was established under the Federal Military Government Decree 32 of 1967 as an idea of two private foundations: Ford and Rockefeller. It became part of the CGIAR in 1971, and is currently governed by an international board of trustees. The Federal Government of Nigeria provided a land grant of 1000 ha. Ibadan is still one of the largest urban centres in West Africa, and is also the home of one of the premier Nigerian universities. There are two principle outstations of IITA in Nigeria: Onne (near Port Harcourt in the south east) and Minjibir (near Kano in the north), as well as others throughout Africa.

IITA has a research mandate for the whole of tropical Africa, and for many years after its founding the research work took place within four programmes, three of which are commodity-based and reflect the early CGIAR mandate (Table 3.9). This list was chosen to minimize overlap with other CGIAR mandates as well as Nigerian NARS. For example, IITA did

Table 3.8 *The CGIAR IARCs and Their Prime Mandates*

Centre	Date of foundation	Date of joining CGIAR	Headquarters	Prime mandate
IRRI	1960	1971	Los Banos, Philippines	rice
CIMMYT	1966	1971	Mexico City, Mexico	maize, wheat, triticale
IITA	1967	1971	Ibadan, Nigeria	maize, rice legumes, root crops
CIAT	1967	1971	Cali, Columbia	cassava, field beans, rice, pasture
ICRISAT	1972	1972	Hyderabad, India	Chickpea, pigeon pea millet, sorghum, groundnut
CIP	1970	1973	Lima, Peru	potato, sweet potato
ILRAD	1973	1973	Nairobi, Kenya	livestock diesease, Trypanosomiasis, Theileriosis
ILCA	1974	1974	Addis Ababa, Ethiopia	livestock production systems, animal feed
IPGRI	1974	1974	Rome, Italy	plant genetics
WARDA	1970	1975	Bouake, Cote d'Ivoire	rice
ICARDA	1975	1975	Aleppo, Syria	farming systems, cereals, legumes, forage
ISNAR	1980	1980	The Hague, Netherlands	NARs
IFPRI	1978	1980	Washington, DC, US	food policy
ICRAF	1977	1991	Nairobi, Kenya	agroforestry
IIMI	1984	1991	Colombo, Sri Lanka	irrigation management
ICLARM	1977	1992	Manila, Philippines	aquatic resource management
INIBAP	1984	1992	Montpellier, France	plantain and bananas
CIFOR	1993	1993	Bogor, Indonesia	forestry
ILRA	1994		merger of ILCA and ILRAD	

not receive a mandate for work on groundnuts due to the long involvement of the Institute for Agricultural Research, based at Samaru, in research on this crop that dated back to initial British interest as an export commodity. In the 1989–2000 strategic plan, IITA was asked to transfer its mandate for rice to WARDA, sweet potato to CIP and cocoyam to a 'competent West African national program' (IITA, 1988). The main emphasis at IITA is now on cassava, maize and cowpea, with smaller programmes looking at yam, plantain and soybean. Throughout its history, decisions over mandates research focuses, for example were taken by the IITA board of trustees with input from various consultants and IITA staff and, in tune with much of the story so far, no resource-poor farmers took part in this process.

Although not part of the Nigerian NARS, IITA does play a major role in the provision of planting material, training and information supply.

Table 3.9 *The Four IITA Research Programmes*

Programmes	Crops
Cereal improvement	maize, rice
Grain legume improvement	cowpea, soybean
Root and tuber improvement banana/plantain	cassava, sweet potato, yam, cocoyam,
Farming systems	all

Increasingly there is also a move towards working with non-government organizations (NGOs). Given the instability with the Nigerian NARS, IITA has played an increasingly important role in the provision of new crop varieties in the country, and like many of the CGIAR centres since the 1970s there have been pressures to widen participation of farmers in setting research agenda. This is a point that will be returned to in Chapter 5. The presence of such an institute within its borders has certainly been of incalculable importance to Nigeria. Indeed Nigeria joined the CGIAR at approximately the same time as it officially adopted the Green Revolution programme (both in 1975).

The first Green Revolution began in Asia during the mid-1960s, and the bedrock was a set of conventionally bred rice and wheat varieties designed to have a good response to artificial fertilizer. This was allied with a shift in US policy away from using its own production to 'stopgap' potentially disruptive food shortages in the developing world and towards the encouragement of agricultural development within those countries (Cleaver, 1972). These crop varieties came from breeding programmes at two stations that later became part of the CGIAR network: CIMMYT (Mexico – focused on wheat) and IRRI (Philippines – focused on rice). The presumption of the researchers was simple – breed varieties that can respond well to inputs and release them to farmers. Increased production was at the heart of this strategy, just as it was with much of the commodity research that predated it. There was much talk at the time about the elimination of hunger and poverty 'at a stroke' (Cleaver, 1972).

The first Green Revolution was undoubtedly successful in the production of more grain. In India, for example, the introduction of high-yielding varieties boosted grain production from 95 million tonnes in 1967–69 to 130 million tonnes in 1980–81 – an increase of 37 per cent (Farmer, 1986). Wheat production alone tripled between 1964 and 1972. All of this was dependent upon an adequate supply of fertilizer and irrigation to give the increased yield, and pesticides to protect it. Associated with these increases in crop yield there were also social, political and economic impacts and some of these were negative, although there are differing opinions with regard to their magnitude (Falcon, 1970; Cleaver,

1972; Ruttan, 1977; Glaeser, 1987; Lipton, 1989; Shiva, 1991; Goldman and Smith, 1995). For example, it has been shown that those best placed to benefit were the richer farmers with access to irrigation and machinery. The poorest could not afford the combination of seed and inputs, and did not have the same access to good irrigation (Lipton, 1989; Conway, 1997). That these impacts were unintended and unforeseen is not surprising given the contemporary practice of focus on technical targets rather than broader socio-economic issues (Farmer, 1986). As we have seen this was a feature of agricultural research in both the British homeland and colonies, and indeed throughout the world.

Following the perceived successes of the Green Revolution in Asia planners began talking about a similar Green Revolution in Africa during the 1970s. The first wave of ADPs sponsored by the World Bank in Nigeria were an attempt to do just that, and coincided with a hike in oil prices that made Nigeria a good place to invest. As already mentioned, the T&V system was adopted and packages of high yielding varieties, fertilizer and pesticide were promoted to the farmers. If rice and wheat yields could be boosted in Asia, why couldn't the same be done for crops such as cassava, maize and millet in Africa? The packages typically revolved around sole cropping at a recommended density based on research results, with fertilizers and pesticides applied at rates known from on-station trials and experience to be biologically effective and leading to increased yields. These packages were derived entirely in isolation from farmers' wishes, and they were seen simply as recipients.

However, despite all this effort and optimism both in Nigeria and elsewhere on the continent (Harrison, 1987), Africa has not experienced the same revolution in grain production as Asia, and indeed the situation could even be said to have worsened with some countries experiencing famine (World Bank, 2000a). It is certainly the case that the modern technologies promoted by the ADPs in Nigeria were simply not taken up to any significant extent in that country. This is not to say that there has not been change – far from it (Goldman and Smith, 1995) – but the type of change, along with its speed and scale, was not that intended by the planners of the Green Revolution. An overriding emphasis on production, backed up by a research base that was isolated and focused on improvement of single crops, did not result in a serious attempt to understand and appreciate the characteristics of agricultural systems in Africa (Lipton, 1988). In the agribusiness of the developed world production was king, and the same was true of all the imperial initiatives, research and infrastructure – to maximize production of cash crops for export. This emphasis on food-crop production also worked with the traditional staples of Asia – wheat and rice. It was assumed that the change induced in Asia could simply be transported to Africa without realizing that higher production of food per se may simply not be the only priority for the

African farmer. Indeed, to this day there is still much emphasis on bridging the productivity gap between research station yields and those typically achieved by African farmers (World Bank, 2000a). Instead, maximizing return and security with the assets available, including land and people, may be the preferred strategy. It is perhaps ironic that some ascribe the many changes in Nigerian agriculture that have occurred since the 1960s to market forces (Goldman and Smith, 1995) – the same commercial forces that drove many of the changes during imperial rule.

SUMMARY AND CONCLUSIONS

In this chapter we have looked at the origins of formal agricultural research in those developing countries that formed part of the British Empire. We have illustrated how this interrelates with the patterns described in Chapter 2. Naturally, given the *raison d'etre* of the Empire, these origins rested heavily upon an emphasis on the production of export crops, and borrowed from the contemporary establishment of agricultural research structures in Britain itself. Research into food crops came later, but still followed the established commodity pattern. Throughout there was a tendency to regard African agricultural systems as unproductive, inferior and requiring improvement. After independence the new states, and Nigeria is an example, inherited both structures and personnel, but unfortunately in many instances turbulence largely resulting from problems of funding and the perceived need to address other pressing priorities set in. The result has been a separation of researchers from those intended to use the outputs of research. Extension was supposed to bridge this gap, but for the most part became a one-way transfer of advice and technology with little or no feedback from farmers to researchers. This in turn has resulted in poor uptake of research results and the failure of initiatives such as the Green Revolution in Africa. The main initiators and implementers of research for subsistence farmers have been the farmers themselves.

One main conclusion arises. If the links between research institutions and those that are meant to use the results of their research are strong then the likelihood is that those charged with the research will respond to the needs of intended beneficiaries. This has been the case with much of the UK agricultural research systems until fairly recently – farmers had their own government ministry and were involved in the establishment and running of many of the research stations. The same is also true of the early commodity stations run by private enterprise or even government throughout the empire. If, on the other hand, research institutions become removed from the users then the danger is that the outputs from the research process will not be applied – simply because it is likely to be inappropriate in the circumstances within which the intended users live.

The implications of this are more serious if resources for research, and indeed for all agricultural intervention initiatives that use the results of research, are limited, which they typically are.

Yet we have also shown that this evolution and its outcomes are understandable given the history, mindsets and pressures that operated throughout. This is not to say that it was 'right' and neither is it to say that there weren't counter-examples that go against this broad picture. A good example of this has been the success of plant breeding programmes, such as that of cassava, going back many years in Africa. Also, this should not be seen purely as an African phenomenon. Much the same sorts of issues of inappropriate research and assumptions as to end-user need have arisen throughout the world, including the developed countries. We would argue that the fundamental driving forces at play are much the same, although their expressed form and function may be quite different.

In a later chapter we look at some of the responses to these problems. They can all be summed up in one word – participation – and are at their heart based on a very simple premise: ask people what they want and use that to set research agendas that are meant to benefit them. Although simply stated, the achievement in practice is not so simple. Indeed, it may be argued that the furore surrounding genetic modification in Britain and much of Europe could perhaps be a manifestation of a complete breakdown of the most vital link of all – the one with the consumer. No-one asked consumers if they wanted GM food prior to the expensive research that took place – it was simply assumed by scientists, the biotechnology companies and politicians that they did, or at least that they were neutral. Not only was there no participation, there are few signs that anyone involved, including those funded by the public purse, saw this as necessary. The results of this, and some other high-profile cases, have been various initiatives to improve the dialogue between science and society. Yet despite the failures of public participation in agricultural research in the UK itself, the major British aid agency DFID, funded by the same public purse, has been heavily involved in promoting the need for end-user participation throughout the developing world, and particularly in its former colonies.

Chapter 4

Modern Times: Agribusiness and Biotechnology

TIME-FRAME: 1947–2000

1947–60	Post-war expansion of agricultural research – 18 new ARC research units and 10 new institutes.
1948	GATT: a treaty to promote international trade and economic development by reducing tariffs and other restrictions. Over 100 countries eventually ratified it, while others apply its provisions de facto. The eighth round of GATT talks, begun in Uruguay in 1986, eventually led to an agreement signed in 1994 to set up the WTO to govern international trade. WTO was subsequently established in 1995.
1950	First generation of entire plants from *in vitro* culture.
1953	James Watson, Francis Crick, Maurice Wilkins and Rosalind Franklin discover the double helix structure of deoxyribonucleic acid (DNA). The DNA transmits heredity information from one generation to the next but it wasn't until 30 years later that even larger strides were made in the field of molecular biology which led to the development of the techniques of modern day biotechnology.
1953	Lord Rothschild pronouncements on agricultural research at Long Ashton.
1958	Creation of EEC.
1960s	Werner Arber's discovery of restriction enzymes which cut DNA strands at precise points.
1961–72	Continued growth of agricultural research – six new units and three new food research institutes.
1961	Civil service classification of research in terms of pure and applied.
1965	The Science and Technology Act – Research Councils becoming a direct responsibility of Government.
1971	The Rothschild Report on 'Organization and Management of Government R&D'.

1972	White Paper – 'Framework for Government Research and Development' – establishment of the Chief Scientist's Group at MAFF and the transfer of some of the science budget to MAFF.
1973	Britain joins EEC. Researchers develop the ability to isolate genes. Specific genes code for specific proteins. Stanley Cohen and Herbert Boyer removed a specific gene from one bacterium and inserted it into another using restriction enzymes – this formed the beginning of the era of recombinant DNA technology, commonly called genetic engineering.
1975	Lome Convention. Trade agreement between EC and 46 African, Caribbean and Pacific Ocean states, aiming for technical cooperation and the provision of development aid. A second agreement was signed in 1979 by a larger group.
1977	Genes from other organisms are transferred into bacteria, including a human gene into *Escherichia* coli.
1980s	Scientists discover how to transfer pieces of genetic information from one organism to another, allowing the expression of desirable traits in the recipient organism.
1982	The first commercial application of this technology is used to develop human insulin for the treatment of diabetes.
1983	Genetically engineered plants resistant to insects, viruses and bacteria are field-tested for the first time.
1990	Publication of the European Directives on the use and voluntary dissemination of genetically modified organisms in the environment.
1994	First authorization by the EU to market a transgenic plant, a tobacco plant resistant to bromoxil.
1995	WTO established as the successor to GATT (see p70).
1996	The EU approves the importation and use of Monsanto's Roundup Ready soybeans in food for people and feed for animals. These beans are genetically modified to tolerate spraying with Roundup for weed control while the beans are growing.

FARMING PRIORITIES FOR UK AGRICULTURAL RESEARCH – DECLINING PARTICIPATION

Many of the early breakthroughs and developments in agricultural research were led by scientists, such as Townshend and Lawes, who were themselves farmers or landowners and felt strong connections with the end-users of their efforts. However, as time went on and communities became increasingly urbanized and distant from the land, agricultural research, too, became institutionalized and somewhat removed from the demands of the farmer. This situation was brought into sharp relief in the post-war years when the requirement for increased agricultural output demanded that research should be there to address the immediate needs of the farmers and yet the link between them appeared to be fragmented.

The Long Ashton Research Station, established near Bristol in the UK as the National Fruit and Cider Institute to study the problems of cider orchards and cider making in 1903, celebrated its jubilee in 1953 and to mark the occasion, Rothschild, the then chairman of the ARC, spoke on the theme of 'Agricultural Research 1953' (Rothschild, 1953). This lecture provides a very interesting assessment of the agricultural research service of the time and contained the following major criticisms:

- Inadequate contact between the research worker and the farmer;
- Inadequate pressure on short-term problems;
- Inadequate knowledge of the order in which these short-term problems should be tackled;
- Inadequate organization of agricultural research (Cooke, 1981).

From this time on, Rothschild became a great proponent of mechanisms to enhance the application of agricultural research. He reinforced the need for accountability of the research council and presided over the setting up of the largest number of targeted research units in the council's history. The post-war years represented a phase of enormous growth in agricultural research and in the number of scientists engaged in this activity. In this period of post-war expansion, until the early 1960s, emphasis was placed by the ARC on fundamental research, and this was carried out largely in specialized units. These were usually set up on the basis of having identified a successful scientist working in a field relevant to the research council's interests, and providing that person with facilities and staff. The unit was usually disbanded on the departure of the lead scientist, through leaving, retirement or death. Indeed, so strongly implemented was the principle of, and emphasis on, units that three were set up during the war, some 16 between 1947 and 1960, and a further six up to 1969. However, one thing had changed irrevocably, and that was the actual involvement or participation of farmers themselves in the research process. The alternative was the politicizing of agricultural research which, by the 1980s, had come to replace participatory involvement completely.

ROTHSCHILD REPORT

Since some of the most significant changes to the ways in which agricultural reasearch is funded in the UK resulted from what is known as the

Rothschild Report, we shall look at it and its implications a little more closely. In this report entitled 'The organization and management of Government R&D' published in 1971, Lord Rothschild, then head of the central policy review staff, listed six questions relating to what might be expected in a review of government R&D:

1 Are we doing too much, too little, or about the right amount of R&D?
2 Is the balance between pure and applied research about right?
3 Should we do more intra-mural and less extra-mural R&D, or vice versa?
4 What R&D are we doing which should not be done?
5 What R&D should we do which is not being done?
6 Is there adequate machinery, at the centre, critically to evaluate the overall R&D scene?

Whilst seeming to us to be very sensible and appropriate questions for the government to be asking, Lord Rothschild essentially dismisses them in the second paragraph, stating that they 'either relate to an out-of-date concept of R&D management, or they are unanswerable'. He goes on 'The resolution of the dilemma, which is more apparent than real, is to ensure that the organization and management of R&D is logical, flexible, humane and decentralized, the prerequisites of an efficient system'. In addressing the six questions, he dismisses the question of balance between pure and applied research claiming it is unanswerable on the grounds that there cannot be a correct balance between the resources devoted to pure and applied research. He also states that the sixth question 'is difficult to answer, because it is doubtful whether any control body can or should try critically to evaluate the government's R&D as a whole'.

These comments illustrate an acceptance that there is control exerted over the direction and management of government R&D; there has to be flexibility, room to respond to shifting needs and opportunities. But in doing so, in accepting this situation, there is a danger that R&D can be led away from the needs of society and perhaps even the needs of the science itself, by strong commercial and political influences. Indeed, the 30 years following this report provide testimony to this.

In the body of the report, Lord Rothschild seeks to establish principles by which R&D can be directed and monitored, with numerous examples out of which come 55 recommendations. Although many of these deal with government organizational matters, a few deal with aspects relevant to the questions we have posed in this book concerned with what drives agricultural research. The first recommendation, and one of the most relevant here, concerns applied R&D:

This report is based on the principle that applied R&D ... must be done on a customer–contractor basis. The customer says what he wants; the contractor does it (if he can); and the customer pays.

Whilst not excluding basic research (as opposed to applied) altogether, the report is clear that the chance observations made during basic research, whilst often providing innovation and opportunity, should not be left to chance and should be concerned with 'the discovery of rational correlations and principles'. The report criticizes the fact that much of the research carried out by the ARC is considered 'applied' but:

...has no customer to commission or approve it. This is wrong. However distinguished, intelligent and practical scientists may be, they cannot be so well qualified to decide what the needs of the nation are, and their priorities, as those responsible for ensuring that those needs are met. This is why applied R&D must have a customer.

That customer, the report goes on to say, should have responsibilities and accountabilities for the R&D programme and should decide how much can be spent in financial terms and other input.

The report tries to make the distinction between applied and basic research, mostly to be able to establish a different set of rules for each with regard to direction, control and monitoring. Interestingly, it recognizes and acknowledges a 'grey' area between the two where applied research touches on pure research, perhaps to answer particular questions relevant to the programme or to pursue some interesting lines of enquiry. The report admits that 'virtually all' applied R&D laboratories will, sooner or later, become involved in such research, either overtly or clandestinely, adding that 'it is a good thing that they do' but requires the activities to be recognized as being necessary and formally quantified. The report calls these (bridging) activities 'general research' and recommends that the input to this area should not exceed an average of about ten per cent of that sanctioned by the customer for the applied research. This 'concession' to the need for pure (basic) research related to applied programmes does, at least, allow for opportunities to provide supporting or underpinning research where this is deemed necessary and/or appropriate.

However, some conflict of understanding seems to have existed at the time of the report. The Council for Scientific Policy did not accept an unambiguous difference between basic and applied research, claiming that the distinction implies a division where none should exist and that this could be harmful. However, the report recommends rejection of the view that there is no logical division between the two, intending to protect the work of the research councils from what it called 'the imaginary ravages of applied R&D users'. However, it recognizes distinctions between the roles of DEFRA and the ARC but says:

> *the present division is both illogical and undesirable. It perpetuates meaningless or harmful distinctions between research and development and panders to scientific snobbery – the 'haves' in the Research Councils and the 'have-nots' in the [DEFRA] Departments.*

Harsh words indeed, though with little effect, it appears, as we consider the same situation today.

On the question of accountability, the report is concerned with establishing a hierarchy, with continuing dialogue amongst the players to ensure effective interaction directed towards the successful outcome of the research. It states that 'without the accountabilities ... both efficiency and the probability of success are reduced'. Accountability in this context rests with the customers who may be in industry or government departments but will, necessarily, be the research programme bosses.

The total estimated expenditure by the five research councils existing at the time of the report, the Agricultural (ARC), Medical (MRC), Natural Environment (NERC), Science (SRC) and Social Science (SSRC), was £109 million. Of this, £18.7 million (17 per cent) was allocated to the ARC with £50.9 million (46 per cent) going to the SRC – between them, the ARC, SRC and NERC accounted for 77.5 per cent of the total R&D budget. However, the report recommended a redistribution of these funds favouring government departments such that the £18.7 million due to the ARC was split with only £4.2 million going to the ARC and £14.5 million to MAFF, and a more or less 50:50 split of NERC's £15.3 million allocation to the research councils and the various government departments, including MAFF.

These recommendations, and the decisions which followed in subsequent years, had far-reaching effects on agricultural research in the UK, not least in helping to reduce the influence and effect of the ARC on how research for agriculture was carried out. This was an attempt to provide a framework on which to base research, its funding and raison d'etre, cutting across some of the bureaucracy and top-heavy management which seemed to have accumulated in the post-war years. Whether it succeeded is doubtful, given the drift away from research directed by needs of society and/or agriculture towards more and more influence from politics and big business. Indeed, a conclusion drawn towards the end of the report states that, in contrast to the view many people have that there should be 'some central body whose duty it is to exercise "general oversight" of government and, even, national R&D', so far as this report is concerned

'...this oversight [for fundamental research] is exercised by the Research Councils...', and for applied research 'the inescapable conclusion ... is that general oversight would serve no useful purpose and, indeed, would negate the principles put forward in the report'. In other words, R&D should find its own level, provide its own direction and monitoring which, again, leads us back to the question – what does drive agricultural research in the UK and what determines funding objectives?

DETERMINING FUNDING OBJECTIVES

The question of what drives agricultural research was raised in Chapter 1. From what was said there, it is concluded that the influences on the driving forces are many and varied – a mix of what science researchers want to do, and where and what funds are available, tempered with input from government policies, industry, public concerns and attitudes and, sometimes, a need. However, without the funding, little research can be carried out at all and so one of the most important factors in determining what drives research is the setting of objectives for funding.

Evidence that policy makers and funding agencies are concerned about determining objectives for directing funding can be seen in the following. As part of their most recent science policy statement, the DEFRA include:

> *In addition to the research which is directly linked to policy objectives, MAFF funds a discrete programme of underpinning science. The intention is still to address Government policy issues, but in areas where the science required is more strategic, cross-cutting or urgent. The aim is to have the flexibility to respond quickly to scientific issues and needs such that important strategic leads are followed up.*

In its most recent mission statement (www.bbsrc.ac.uk), the BBSRC states that it is established by Royal Charter for three purposes:

> 1   *to promote and support high-quality basic, strategic and applied research and related postgraduate training relating to the understanding and exploitation of biological systems;*
> 2   *to advance knowledge and technology, and provide trained scientists and engineers, which meet the needs of users and beneficiaries (including the agriculture, bioprocessing, chemical, food, health care, pharmaceutical and other biotechnologically related industries), thereby contributing to the economic competitiveness of the United Kingdom and the quality of life;*

3 *to provide advice, disseminate knowledge, and promote
public understanding in the fields of biotechnology and the
biological sciences.*

There is surprisingly little mention of agriculture here, and perhaps even less that might be of direct benefit to agriculture – perhaps agriculture is no longer intended to benefit from the outputs of DEFRA and BBSRC's research? We comment further on this later.

In considering the beneficiaries of research in this context, significant changes can be noted over the years. In times of great need, such as during periods of war, famine or depression, research has been directed towards addressing more immediate problems of food production. In tracing developments since World War 2 we can see considerable change in the attitudes and focus of the agricultural research councils and in the type of research being carried out. During and just after the war, expertise was directed towards agricultural production and research objectives were very practical and applied, with little scope in either funding or time and effort available for more fundamental basic research. With increasing prosperity in the following years, we have seen a significant change in the remit of research councils away from the applied function and reference to agriculture towards more fundamental work, leaving the applied research to be picked up by industry. We return to this question towards the end of the chapter.

GUIDANCE FROM FORESIGHT

The UK Foresight Programme was launched by the government in 1994 following a major review of government science, engineering and technology policy. In 1995 the first set of recommendations for action were published, followed by four years of development and implementation. A new round of the programme began on 1 April 1999, managed by the Office of Science and Technology in the Department of Trade and Industry. The results are:

- being used by companies, large and small to reshape their business strategies and build sustained competitive advantage;
- breaking down barriers to collaboration across business sectors and academic disciplines and between business and the science base;
- focusing business and its science base on key issues for the quality of life and informing policy and spending decisions across government.

At the heart of the Foresight Programme are panels which work in each of the following 15 sectors:

- agriculture
- natural resources and environment
- health and life science
- chemicals
- information technology (IT) and electronics
- communications
- construction
- defence and aerospace
- manufacturing, production and business processes
- materials
- energy
- retail and distribution
- financial services
- transport
- food and drink

They bring together representatives from business, the science base, the voluntary sector and government to consider the future and make recommendations for action. Each panel is supported by a number of task forces which look in more detail at specific areas. There are two kinds of panel:

1   thematic panels which address broad social and/or economic issues which might drive wealth creation and affect quality of life in the future; and
2   sectoral panels which focus on business sectors and broad areas of activity and carry forward the work of existing panels, as well as tackling new issues.

The Foresight Programme is supported by associate programmes undertaken by other organizations, mainly professional institutions and research and technology organizations. Associate programmes investigate the future of a particular topic, within the framework of the national programme. The result of panel, associate programme and other activities are fed into the knowledge pool. The process involves a flow diagram starting with social, technological, economic, environmental and political changes leading to the identification of global and UK needs, opportunities, and threats, and of the challenges that are being faced in defining the knowledge and resources required for effective responses, and the actions for wealth creation and improved quality of life.

The Advisory Council for Applied Research and Development (ACARD) (2000) published the results of a study on 'promising areas of science' – the objective of which was 'to survey current scientific developments, review progress in the UK since 1986 and advise the Council on work which showed commercial and economic promise in the medium to long term.' The report entitled 'Exploitable Areas of Science' (HMSO, 1993) concluded that:

> ... *a process is needed to prioritize and guide a substantial proportion of that part of the National Scientific Resource, be it Research Councils, Ministry of Defence or Department of Trade and Industry, and to stimulate its effective exploitation to the benefit of the UK.*

It went on to state:

> *We do not have a forum in the UK where we can manage [this] process... It is, we believe, a matter of national priority that such a forum be established.*

The report outlined a framework for the process of generating strategic research priorities. Among the questions to be addressed were:

- Which areas of generic technology are supported by a particular area of strategic science, or for what areas of basic science can applications be found to support them?
- Does the UK have the scientific resources to advance a particular area of strategic research?
- For a given generic technology, what new products or processes will become possible within 10–20 years?
- What are the likely costs of translating scientific knowledge into marketable products and processes?

The report also identified four keys elements in the process of identifying exploitable areas of science:

1 the gathering of information on a continuing and permanent basis and its communication to the relevant parties and bodies;
2 the evaluation of relevant opinions and information, and the identification of exploitable scientific areas;
3 the allocation of resources to the priority areas in science;
4 the commitment to exploit the results of science to UK benefit.

The report noted that 'the information necessary to identify exploitable areas of science is acquired at present in a fragmented fashion in the UK.' A number of bodies such as ACARD, the Royal Society and University Grants Council (UGC), together with industry, all play a role but rarely do they interact as a combined force to shape policy and direction. A structure is required which can gather, analyse, prioritize, and direct relevant information to the decision-making machinery.

There are a number of points which arise. First, it is interesting to note that ACARD saw fit to investigate the 'promising areas of science' and provide a review of progress. Their basis for selection of these 'promising areas' appears to have been those showing 'commercial and economic promise in the medium to long term' – in other words, selection appears to be have been purely in a business sense, for the benefit of commerce and profit rather than society or mankind in a more altruistic sense (and no, they are not the same thing to most people not involved in business). The report expands the remit 'to prioritize and guide a substantial proportion of that part of the national resource... and to stimulate its effective exploitation to the benefit of the UK' – that is, 'UK plc', considered as a nation of keepers of businesses somewhat larger than the average corner shop. Again, does this really benefit society as a whole, does it address specific problems that people may encounter every day? 'We have no forum where we can manage [this] process...' but that forum is, we suggest, very unlikely to achieve much beyond larger profits for already large concerns – that is, after all, the report's stated aim. Throughout, the report is concerned with identifying opportunities for 'marketable products and processes', the production of goods, products and procedures for sale to create wealth from exploitable science in the strictly commercial sense.

Whilst we must always acknowledge the need for returns on the investment in science, the report provides us with an excellent example of how attitudes during the greedy 1980s and early 1990s tipped the balance in favour of commercial exploitation of research in which its worth was measured purely on the basis of its commercial value to industry. This was often at the expense of more fundamental research, the more innovative and speculative research which provides the platform for the next bandwagon, that which evolves from tangential sparks of ideas that need fanning with funding to provide the flames of tomorrow.

The report concludes that, in the UK, there are certainly interesting examples of Research Foresight going on, but these are carried out in isolation from one another: there is no interaction at different levels. There is also a strong, national focus in all this, directed almost entirely at economic growth and commercial advantage to the UK with very little attention paid to international interaction, the needs of society or mankind in an altruistic sense; it is isolationist and profit-driven. This is not the case elsewhere. For example, in Japan, Germany and the US, similar exercises take place within companies, within industrial associations, within individual government departments or funding agencies and span all research. Furthermore, there is some integration between the different levels. In Britain, there is no such integration. In particular, there has been no attempt so far to produce a holistic overview of science and technology. Yet without this, it is impossible to analyse the interactions between different technologies – that is, the potential for technology fusion. There is also too much reliance on the private sector and there is no national forum for identifying priority areas of exploitable science in a systematic manner. ACARD's aspirations of 1986 remain largely unfulfilled according to this report.

CONTRIBUTIONS FROM THE EU

It is interesting to note the evolution of the EU Framework Programmes. The earlier forms up to Framework IV in the late-1990s were based very firmly in the scientific research community, addressing problems that were essentially scientific or technical with little or no consideration for the broader application, although, interestingly, rather less so for the agriculture programmes. In the most recent Framework Programmes, this has changed significantly to become very much more orientated towards using science and technology to solve problems in society and in the community. But this still allows some of the more fundamental research by encouraging multi-disciplinary research groupings involving a range of scientific expertise focused on specific objectives.

The EU's Fifth Framework Programme (FP5) for research, technological development and demonstration (RTD) was agreed by the Council of Ministers of member states of the EU, with the European Parliament on 13th December 1998. The activities covered encompass all research in the EU and aim at improving the 'competitiveness' of European industry and the 'quality of life' for its citizens. FP5 also provides scientific and technical support for other common policies. There are significant advantages to pooling resources and knowledge through collaboration at a European level, rather than individual member states acting on their own. EU RTD complements national research efforts and the European community as a whole reaps the benefit.

The research priorities of EU FP5 have been established on the basis of criteria divided into three categories:

1   those related to the EU 'value added' and subsidiary principles (so as to select only those objectives which are more efficiently pursued at the EU level);
2   those related to social objectives (in order to further major social objectives of the EU reflecting the expectations and concerns of its citizens); and
3   those related to economic, development, scientific and technological prospects (in order to contribute to the harmonious and sustainable development of the EU as a whole), with additional criteria being established by specific programmes.

(see www.cordis.lu)

Competitiveness and quality of life are terms which have evolved with the EU programmes and have taken on meanings of their own. Indeed, the EU programme writers, planners and the EC are obsessed with improving Europe's competitiveness, presumably intending to mean the ability of the EU as a trade group to compete on the world market. Improving the quality of life is a phrase used frequently in EU literature but it is a vague term with many likely interpretations. It is hoped it is intended to apply to everyone by raising living standards and providing improved opportunities for all. However, these terms are loaded with political meaning and generalizations which require careful interpretation in whatever context they appear.

POLITICAL AGENDA OF EU INTEGRATION?

FP5 is described as differing considerably from its predecessors, having been conceived to help solve problems and to respond to the major socio-economic challenges facing Europe. Its impact is maximized through focusing on a limited number of research areas combining technological, industrial, economic, social and cultural aspects. In this way, topics of great concern to the EC can be targeted by mobilizing a wide range of scientific and technological disciplines, both fundamental and applied, which are required to address specific problems, overcoming barriers amongst disciplines, programmes and organizations. FP5 is directed mostly at strategic rather than near-market research (that is, basic research targeted towards applications rather than applied research per se) but there is an increasing emphasis on maximizing exploitation of research results and the transfer of technology into the market place. As a result, projects are expected to demonstrate a clear strategy for the subsequent development of their results into new marketable products, better manufacturing processes or other means of exploitation or dissemination. This is a clear indication that the EU sees the need to maximize the returns of its funding as part of its remit.

APPLYING APPLIED SCIENCE

In similar vein the UK LINK Collaborative Research Programme extols its advantages to participants by listing the benefits to both industry and academic researchers. Benefits to industry include:

- Companies having access to high quality research and leading-edge science which can underpin business strategy and innovation;
- As partners in government's principal mechanism for supporting collaborative research, companies have the opportunity to 'shape the future of the UK';

- LINK accelerates the research process and helps reduce the investment required for each industry partner;
- LINK undertakes research that is non-competitive and involves an element of risk which would not be carried out without government support;
- It provides partners with means to create new intellectual property and new improved products, processes and services;
- It encourages the use of an extensive and world-class research base to increase UK industrial competitiveness;
- It provides excellent opportunities for networking and sharing ideas with experts from many fields;
- Its projects are focused on innovative ideas which are likely to stimulate publicity, and raise the profile of a company in the national and international arena and help extend the company's horizons.

Similarly, benefits to academic researchers listed include:

- The research base partners gain significant additional funding from the government and industry;
- Collaboration with industry increases the scope and facilities available for research;
- LINK projects often lead to new avenues of research and technology transfer, and offer regular opportunities to exchange views with other disciplines, research establishments and industry experts;
- Projects can offer rewarding experience for academic researchers in working with industry and it enables them to sample research and career opportunities in an industrial environment;
- The research base plays a key role in the innovation process and provides the opportunity to engage fully in this process and contributes to the health and prosperity of all.

Further concerns to maximize the benefits of investment in research can be seen in a report in which John Baker (1999) states:

> *I have been asked by the Treasury and DTI ministers to investigate the commercialization of research in the government's public sector research establishments, focusing in particular on issues of good practice, barriers to successful commercialization, culture, management and the public sector research establishment's sponsor body relationship, and to make recommendations for increasing the rate at which their research is successfully commercialized, consistent with other government objectives for public sector research establishments. This is with particular reference to the role of*

*sponsored departments/research councils in promoting the
exploitation of research in institutes; to making progress in
improving the culture of entrepreneurship within research
institutes; consideration of the organizational capacity and
expertise for managing and exploiting government intellectual
property effectively; considering specific institutional barriers
and possible new incentives, and the scope for closer co-
operation with the private sector; and spreading best practice.*

The report comments that the current state of play is that many government
research institutes are involved in commercializing their research and
expertise through collaboration with industry to solve problems (often in
the context of contract research for industry), and the licensing of
technology to the industry user, either directly or through intermediaries.
Many are also engaged in the sale of services, data and software to the
business sector. In some instances, public-sector research establishments see
the free dissemination of their research outputs as the most effective means
of knowledge transfer, with economic benefits accruing to an industry as a
whole, rather than to individual players.

There is little, if any, mention of agriculture or agricultural research in
this document. Furthermore, agricultural research seems to receive low
priority in a report to the Office of Science and Technology by Policy
Research in Engineering, Science and Technology (PREST) based at the
University of Manchester, published in September 1995. The report
presents the results of the Delphi Survey carried out as part of the UK
Technology Foresight Programme and designed:

*to allow the fifteen panels of experts to consult widely in the
business and science and technology communities, to assist in
achievement of commitment to results and consensus on
developments, and to inform these communities about the
major issues being addressed in the Programme and how their
peers assess those issues.*

Each panel identified topics for consultation, reflecting the market
orientation of the UK programme. Interestingly, and somewhat significantly
perhaps, agricultural research hardly gets a mention in the report summary
although we can deduce something of current attitudes when we examine
some of the detailed findings.

In the chapter devoted to the detailed analysis of agriculture, natural
resources and environment the report presents lists of topics considered by
the respondents to be most and least important, taking various criteria into
account but with the overriding consideration of their impact on wealth
creation and quality of life. The report reveals that, for agriculture, the

responses to many topics are typified by the dominance of wealth creation at the expense of quality of life, or vice versa, with few having equal weight, and the absence of topics where these are seen to be favourable or unfavourable simultaneously is notable. The idea of sustainability underlies many of the topics, along with development relating to food products, animal health, the social aspects of agriculture, production and machinery, and trees and wood products. Practical topics include agricultural productivity, often involving genetics, more effective extraction and use of natural resources, and improving management of the environment.

Amongst the top ten topics relating to agriculture, as judged by their likely beneficial impact, are:

- development of improved genetic engineering technologies to produce new industrial products in livestock;
- increased quality and value of food, feedstuffs and non-food products and the development of non-allergenic foods;
- better understanding of immunology and epidemiology of disease processes in animals;
- widespread public acceptance that the new technologies emerging from modern biology lead to significant wealth creation in the UK.

Amongst the bottom ten are:

- increased knowledge of animal preferences and causes of stress;
- practical use of techniques for multiple embryo production (clones) from genetically superior livestock;
- widespread use of herbicide-resistant crop varieties;
- practical use of molecular probes and genetic fingerprinting techniques to determine the genetic origins of plant and animals for the purpose of protecting biotechnological investment.

Not surprisingly, perhaps, given the current disastrous situation with the prevalence of BSE and the foot and mouth outbreak, the top-ten list is dominated by concerns for improvements in food and feedstuffs quality, and the understanding of animal-related problems, including new technologies. However, it is not easy to recognize any of the topics in the more traditional sense of agriculture and farmers' needs, indicating the extent to which current agricultural research has moved away from farmer participation.

ACADEMIA AND INDUSTRY INTERACTIONS

A study from policy Research in Engineering, Science and Technology (PREST) (Cox et al, 2000) examines the links between industry and higher education institutions HEIs) in the UK. The report states that there has been:

> *spectacular growth in recent years across the UK in the scale, number and variety of linkages between HEIs and industry...manifested in research collaboration, provision of consultancy services, market transactions in the commercialization of research.*

Funding of research by industry grew by 30 per cent in the three years to 1998. Access to industry funding is considered to be a strongly motivating factor by HEIs but only as a means to pursue objectives acceptable to both the academic and industrial partners. The requirements for collaboration are seen as 'mutual trust and a professional, business-like approach by the academic partners'. The authors of the report saw it necessary to add: 'Motivating individual academics to work more with industry requires an incentive structure of similar weight to that of the Research Assessment Exercise' (RAE). The securing of intellectual property rights (IPR) in the process is also seen as important, especially in the commercialization of academic results, but there is concern about the cost of exploiting and protecting IPR. Barriers to academic-industrial research collaborations remain, often arising out of the different objectives and aspirations of the two types of organization and their associated scientific researchers and, in some cases, through conflicting views about the RAE in relation to industrial collaborations. The most serious effect of this is the increased pressure the RAE places on academics to publish their work – more often than not the performance of academic researchers is judged on the basis of their publication record and publication-oriented activities which, inevitably, displaces time available for the more applied industrial research and the development of industrial links, let alone teaching. There is little or no encouragement or incentive in the RAE in practice to pursue such links – if policy-makers wish to promote the full range of industrially relevant activities, then more attention and emphasis must be placed on fostering these academic–industry links with due recognition and reward.

This requires a significant shift away from the currently accepted measures of performance of academic researchers (perhaps better described as researchers in academic institutions) which are heavily biased towards publications in the highest ranking journals (see Chapter 1). The current system is fraught with difficulty and loathed by all except those who benefit, such as those academic researchers involved in esoteric high-powered research in trendy areas of science who can publish their results in high-ranking journals and readily ignore the need to make their work 'relevant to industry', and those who require a 'score' for assessing such work to rank their performance against others. It leads to undue emphasis being placed on academic work published in academic journals; the work is usually short term and is rarely applied or strategic. The system penalizes researchers who wish to apply their efforts to interacting with industry in more strategic areas where the goals are usually set by the industrial partner(s) and are therefore usually directed towards a product, procedure or tangible end result of some sort. Publishing results is not ruled out but usually has to take second place and may be delayed for confidentiality reasons. Furthermore, it is mostly the case that the journals best suited to receive such work do not rank very highly in the 'merit' stakes. It should also be borne in mind that this system detracts from academics' ability to provide time and energy for educating their pupils. It is an unfortunate and ridiculous system which has evolved during an era where everything has to be accountable with bureaucrats devising this measure of scientists' performance; it can be likened to the methods of assessing driving ability by monitoring speed – simple and seemingly effective but subject to considerable abuse on both sides. The current system of assessing scientists' performance also heavily penalizes long-term research by placing further undue emphasis on the number and frequency of publications as part of the researcher's 'measured output'. A 'score' is arrived at by multiplying the number of publications by the rank value of the journal; this exacerbates the unfairness to the point of farce when comparing all researchers against one another – the sad and serious point of this is that the system is applied, the results are believed and used to determine future strategies and the comparative performance of research output. It is a wonder any applied and strategic research is carried out at all, let alone in collaboration with, and subject to, industry-led objectives, given that the assessment system favours so called 'high performers' in this way.

FARMER PARTICIPATION?

Maybe we can take some comfort from the fact that the situation has spawned surveys and committees to monitor and review the whole process, and numerous reports and publications (all, no doubt, in high-ranking journals!) in various guises examining the relationships between academic research and industry, with incentive and initiative schemes like LINK and Foresight and others designed to encourage industry–academic links and the exploitation of the fruits of research. This has had the desired effect in some areas by providing assistance in the form of the all-essential funding and, in schemes like LINK, by providing programme and project managers to oversee the work and monitor progress and fair play amongst the participants. But scientists, and those in agricultural research often have to make a conscious decision between a career in academia – competing on the basis of publications in high-ranking journals and international recognition of their work – and a career interacting with their industrial counterparts, driven by the satisfaction and rewards of achieving commercially-oriented objectives of the projects.

These two forms of research dominate the scene today with little or no attention paid to 'mud on boots' farmer-driven research. This is partly due to the above in which the 'measures of performance' are biased towards the trendy high-tech science at the expense of the more mundane, albeit valuable research, but also because modern-day research is seen as expensive and researchers seeking funding will be attracted to those areas to which the funding is directed. It is ironic that the seemingly more mundane research which can contribute enormously to our basic understanding of situations in relatively simplistic but important ways, is usually very inexpensive compared to the high-tech work – but it is not fashionable and very few funding agencies will entertain the idea of supporting such work.

One of the barriers to progress in this direction is the seemingly innocuous term 'high quality science'. For many people, and unfortunately most of those concerned with the allocation and justification of funds, high quality science is synonymous with high technology and so called cutting-edge science – anything else is less – which, of course, simply excludes all those valuable, constructive but simple projects which require no more than basic equipment to provide useful data and information. This is, of course, an oversimplification – there are few experiments these days which cannot be made more comprehensive and productive through the judicious use of modern day technology – although one is reminded of the quote from the science fiction writer Paul Anderson: 'I have yet to see any problem, however complicated, which, when looked at in the right way, did not become more complicated!' But the principle remains – research involving the participation of farmers nearly always and inevitably suffers at the hand of high technology these days.

So, is modern-day agriculture not sufficiently high-tech to benefit from this trend? Of course it is, but other factors come into play here – those which have changed farming into a high-tech business which is profit-driven so that any research that is directed towards solving its problems must satisfy the 'profit god' first and foremost. One obvious exception to this, emerging as a direct result of the backlash to some recent catastrophic failures of high-tech science, is the move towards organic farming.

A specific example of the complexities of the interaction between the players in these games – the public, politics and industry – can be taken from a scientific paper (Joly, 2000) which considers the various influences on the GM debate. The paper analyses the failure of the introduction of GM crops in Europe and challenges the idea that this failure is due to the influence of public irrationality on decision processes and that policy makers appear not to be willing to base their decisions just upon 'sound science' but to include other elements in the picture, resulting in so-called 'biopolitics'. This is a complex process which simultaneously operates within a public attitude and a scientific agenda. Some might view this as a demagogic way of dealing with problems but the paper challenges this idea and portrays it as the experimentation of a new model of technological democracy. In putting forward its case, the paper addresses three questions:

1 that the claim of 'irrationality of the public' is not based on solid ground;
2 that 'sound science' has limitations in the context of decision making related to uncertainties and controversies; and
3 the requirements for the new organizational and institutional devices in order to cope with these problems, leading to the debate on precautionary principles.

Various surveys carried out to assess the situation as the GM debate evolved have identified that 'perception of risk' is a marginal component of public support or criticism for biotechnology. The decrease in public support is not correlated with the perception of risk but is highly correlated with a decrease in perception of usefulness, and the moral implications of particular applications. Ethical concerns are not separate from the public appraisal of GM science, with serious concerns about the speed of change and that too little time has been taken to evaluate the social desirability and usefulness of developments and their social and human consequences. Furthermore, scientists who do not express doubts about the process may be considered arrogant. The issue of 'usefulness' refers to a wider assessment of the agricultural production model which points to the wide variety of food products already available and the existence of food surpluses, combined with the adverse effects that the highly intensive agriculture normally associated with biotechnological developments will have on small farmers and the environment. 'Usefulness' is also identified with the pressure on public decision-making related to commercial interests and perceived long-term uncertainties over risks and consequences. Furthermore, there is a strong feeling that GM crops should offer distinct, measurable advantages over conventional crops as, for example, in the use of an insect-resistant crop which is judged 'unlikely to cause more harm than a conventional crop sprayed with chemical pesticides'. The argument is that GM crops should offer environmental improvement over conventional agriculture. Such views are likely to become more prominent as international agreements concerning sustainable development begin to permeate policy more widely.

The paper concludes that the way in which GM crops were introduced in European countries appears as a major failure of the institutional systems, and is likely to result in the commercial use of GM crops being blocked for many years. However, the paper states that the situation is not the result of the irrationality of the public but the inability of the various institutions to accommodate the deep concerns expressed by the wide range of players, including scientists, politicians and the public. The solution will be in a major renewal of the framework of public policy, based on precautionary approaches and wider public participation.

Seen in the context of the funding of agricultural research and the questions we have posed, this paper raises some interesting points relating to what has become a hotly debated issue – that of the development and introduction of GM crops. It provides a good example of just how effective public opinion can be in influencing the direction and outcome of modern day research, through the new 'biopolitics' and technology democracy. The introduction of GM crops in Europe was driven strongly by commercial interests – not in the sense of national wealth creation but by a few very large and influential multinational companies who saw clear opportunities to gain much control and profit from this application of GM technology. The justification for the need was couched in terms of feeding a starving world and requiring significantly lowered inputs of agrochemicals with little mention of potential profits and farmer obligation to the companies, all of which come with the new technologies. The forces driving the research which gave rise to most GM products have been very clearly commercial and profit-based.

FUTURE – WHAT FUTURE?

In all, 55 recommendations were made in the 1971 Rothschild Report and although many of them concerned organizational and responsibility matters, a number gave rise to decisions which shaped the future of agricultural research in the UK for the next 20 years, with increasing tendencies towards political influences and those of the large, multinational companies. The report firmly established the customer–contractor principle, with ever-increasing emphasis on accountability and value for money. Another consequence of these changes is the move towards short-termism, the use of short-term contracts to carry out short-term research. Indeed, long-term views in research have become unfashionable for the large part and, more seriously, have undermined long-term strategic (in the true sense of the word) approaches to long-term problems. This has led to the demise of research associated with practically everything that cannot be resolved within a three to five-year contract period. It is most unfortunate for those problems requiring longer time spans of work that this trend towards highly specific short-term contracts has coincided with the rise of biotechnology and a raft of techniques which can speed up many of the more traditional time-consuming procedures. This places greater emphasis on the need to get a whole host of things done quickly, within the three-year time span, made possible (just) by concentrating on what fits into this restricted time scale rather than considering other, perhaps equally or even more important but less accessible longer-term research projects. Why should funding agencies consider funding in blocks of more than three years when there seems so much to fund in this category anyway? But, make no mistake, it has all been at the expense of the long-term, strategic view and, more crucially, at the great expense of many scientific careers.

Concerns over the funding of agricultural research in the UK are expressed in the concluding remarks in the chapter on 'Agricultural and Food R&D and Technology' in the excellent and comprehensive report entitled 'Science and Technology in the UK' (Cunningham, 1998). The 1971 Rothschild Report is put into the context of providing a means of closing the gap between the largely applied research supported by MAFF and the mostly basic research carried out within the BBSRC through its customer–contractor principle. However, political dogma during the 1980s, forcing the idea of competitive markets as the key to economic success, began to transform the organization of public funding for research much more radically, identifying three major effects of this political influence on the organization of food and agricultural research:

1   Total public expenditure was reduced by 7 per cent and core funding by 13 per cent, assuming that the deficits would be filled by savings resulting from closing or amalgamating research institutes or selling them to the private sector, and by charging for services previously provided free of charge, and, further, by encouraging levy-funded research in which farmer and producer profits are 'top sliced' to create funds for research projects.
2   Research towards improved productivity and near-market research, including technology transfer, were made the responsibility of the private sector, so that public funding could be directed towards basic research. This resulted in the abolition of the AFRC in 1994, and some of its activities were transferred to the newly formed BBSRC which became increasingly fundamental in its research remit.
3   Publicly-funded research is increasingly being carried out in higher education establishments which bid for contracts from DEFRA, BBSRC and DTI, for example, competitively.

However, DEFRA has been forced to dramatically increase its funding of food-related research in recent years following public concerns over food safety after several national incidents – several food poisoning cases, the deepening BSE crisis, foot and mouth and the ongoing GM crops debate.

This increase in funding has more than compensated for the cuts in near-market research but at a very high cost to public confidence.

The situation has been exacerbated by the private sector not responding to fill the gaps left by the withdrawal of funding directed at increased productivity and near-market research to the extent expected. Although funding for food research has even increased to address the various crises, there is no evidence that cuts in funds for agricultural research have been made up and so this area has suffered a steady decrease in real terms in recent years. The rationale behind the cuts gambled on increased input from the private sector to plug the gaps in funding, especially in the areas of near-market research and technology transfer, but this has not happened to the extent anticipated by those enforcing the cuts. This has been blamed on the sluggish economy of the 1980s and 1990s but it may also be that the new biotechnologies, genetic engineering and molecular-marker technologies, have not been sufficiently advanced towards market needs to attract investment and support from industry. There are signs that this is changing – certainly the new technologies, their advantages and applications, are strongly promoted by BBSRC and its institutes – but this change is tempered by significant loss of public confidence in science and technology following the scares of recent years. The effect is stalemate, and the future remains unclear.

Cunningham (1998) concludes with the remarks:

> *In summary, we have found strong evidence of market failure and the need for continued and increased public funding of agricultural research. Yet critics of the reforms of agricultural research over the past 25 years have noted that neither agriculture nor food appears in the title of the newest public body responsible for research relevant to these industries, the BBSRC, and that MAFF is no longer involved in promoting the more efficient production of food on British farms. This together with the reluctance of the private sector to fill the gap, make it difficult to disagree with Spedding (1984) that there no longer is publicly-funded agricultural research in the UK.*

It is ironic to read in the book 'Agricultural Research 1931–1981', published in 1981 under the auspices of the ARC to celebrate the success of British agriculture, that 'One of the very strong components in the present agricultural research service is the participation of farmers and growers…'.

The final remarks state clearly: 'The government needs, therefore, to reconsider its policies to better ensure that the public and private sector produce a sounder allocation of resources'. The food industry, and the agricultural industry essential to sustain and promote its supply of high quality, safe produce, appears to be increasingly unsupported by relevant and appropriate research with the continued withdrawal of funding and significant changes in political attitudes and dogma towards funding of underpinning, basic and supportive research.

Along with the decisions which determine where the funding will be targeted comes the need to ensure that it is spent wisely and effectively. This need will depend on the type of research being carried out, who is paying, who is doing the accounting and to whom the accountability is addressed.

Chapter 5

The Rise (and Further Rise?) of Participation

INTRODUCTION

Previous chapters have illustrated how a 'top-down' emphasis in agricultural research arose in both the UK and its colonies as a natural consequence of the circumstances prevalent in each. In the UK it was an increasing desire of government to intervene so as to ensure that food security was maintained in the event of war. Even so, the farmer interest in guiding such research was strong and farmers had a major influence in policy. In the latter part of the 20th century this balance changed with the rise of agribusiness and a move towards biotechnology. The power of the farmer to direct public-funded research agendas has diminished, although there are a number of privately financed research stations. Instead the agendas are increasingly being set by government agencies and business, with little or no farmer, or indeed consumer, representation. The research environment has increasingly become highly competitive, with UK-based researchers operating within a global market.

In the British colonies the history of agricultural research was quite different. A top-down emphasis in the colonies arose mainly as a consequence of a focus on cash crops for export rather than subsistence systems. Also few people involved in directing the research agenda actually came from those countries, or if they did they had been educated and trained in Britain. Therefore there was no link between subsistence farmers and those that drove the research agenda. An additional factor that flowed from this surrounded the 'valuation of knowledge'. The result has been an emphasis in many extension programmes on the transfer of technology (TOT) approach: valued knowledge generated by the formal research community is passed on to the farmers. Given that much of the technology generated in this way was inappropriate, uptake has generally been poor. It could be argued that the major agricultural impacts of European imperialism in Africa have been in terms of helping to facilitate indigenous-led agricultural change, much being inadvertent rather than planned.

This chapter first describes the responses of those that attempted to tilt the balance away from a top down and TOT emphasis to one of farmer inclusion and participation. Since the late-1970s there has been an increasing emphasis on participatory approaches in development, and particularly in the guidance of agricultural research programmes. The participatory movement quickly won the philosophical debate and the emphasis has turned towards implementation. Yet even after 20 years there are problems. This point is discussed more fully in Chapter 6.

In this chapter we briefly explore the history of participation and the number of strands that gave rise to it in agricultural research within the developing world. Only a few of these can be covered, and the intention is more to display the underlying concerns and diversity of approach rather than provide an exhaustive review. For example, no mention is made here of Soft Systems Methodology (SSM) or action research (AR). This is not to deny their importance and relevance. Links will be made back to developments in setting agricultural research priorities in the developed world, primarily because it is our thesis that the two trends, although apparently separate and often treated as such in the literature, have influenced each other. To better understand the evolution of participation in agricultural research in the developing world one also has to understand what has happened and why in the developed world. The two are not separate, but inextricably entwined.

As with Chapter 3, the emphasis here is mostly with the formal research systems established and primarily funded by government and other donors as distinct from indigenous or informal research. We in no way wish to imply that the latter is not important or somehow plays a secondary role to the formal research sector; if anything the opposite is the case.

THE ORIGINS OF PARTICIPATION

The notion that the ultimate beneficiaries of any agricultural research programme should have an input into the direction of the programme is, of course, logical and is certainly not new. As we have already seen, this was effectively the 'default' position when many of the agricultural research stations were created in Europe and North America, and guidelines that strongly promote the importance of seeking involvement from potential users exists to this day (Porter and Prysor-Jones, 1997). Even in the colonial period, given that the initial emphasis was very much upon cash crops for export it may be argued that the mandate of many of the early research stations did indeed match those of potential users – the plantation owners. As already mentioned in Chapter 3 many of the colonial governors, the individuals with the ultimate administrative authority within the colonies, were rich landowners and farmers. No doubt there were individual farmers and plantation owners that felt their views did not receive the attention they deserved – it is almost impossible to please everyone given that resources are always limited – but by and large participation of beneficiaries in directing research was the norm rather than the exception.

Ironically, the logical desirability of ensuring that agricultural research matched the needs of those it was intended to serve broke down in the latter half of the 20th century in both developed and developing countries for quite different reasons. As we have seen, in developed countries such as the UK there was an increasing trend towards commercialization of agriculture and a rise in the power of retailers of produce and a concomitant decline in the power of the producers. In parallel, many of the stations transferred to government control and mandates for research became increasingly set by government bodies – albeit with scientists and other interest groups as advisors. The potential users were no longer just the producers, but became a diverse group with quite different degrees of power and influence. The research base also diversified to take account of this, and the result has been a complex mix of agendas and pressures geared towards a diversity of potential users. In some cases the potential users of the research may not be entirely clear, and government policy often encouraged this! In the 1980s, for example, the UK saw a period of emphasis in its public-funded research away from applied research, which the government felt should be funded by private interests, towards what was termed 'blue sky' objectives. The latter may not necessarily have had a single potential user immediately in mind, and may have stressed more multiple benefits towards society as a whole including, for example, environmental benefits for consumers, producers as well as retailers. In the developed world potential user has often become synonymous with funder.

In the developing world the dynamic was quite different. A legacy of cash-crop-centred research and the physical infrastructure that embodied it did not resonate with the needs of the vast majority of the farmers in those countries. They were not rich, and neither did they have any history of political power within the 'created' countries in which they lived. While it is true that under the British colonial regime there was an increasing emphasis on staple food-crop research, this was all implemented within the research station system that had served cash crops so well. Given the complexity of tropical agriculture and rural livelihoods, it is perhaps no surprise that early efforts often missed the real targets. This became increasingly apparent throughout the 1960s and 1970s, and the result in the developing world was a rise in participation. This was a phenomenon of the developed world's interaction in terms of aid to the developing world. As such the major players in this movement were scientists, mostly social, and academics in the developed world. The movement did not arise from a primarily indigenous desire, although as already mentioned the vast bulk of the agricultural research in the developing world was carried out by the farmers on their own farms. The main features of the rise of participation was an emphasis on methods and mindsets that allowed the voice of the rural people to be heard by those charged with developing technologies in the research sector.

Therefore, the agricultural research trends in the developed and developing worlds were in many ways in opposite directions. In the developed world there was a move away from a simple focus on the producer to addressing the concerns of a multiplicity of potential users, each with quite differing, and often conflicting, needs and power. This has been in part a result of complex policy shifts related to political pressures, but also a function of power related to financial power. In the developing world there was a shift in emphasis from one small powerful set of producers to a much broader, yet typically far less powerful, set of resource-poor producers.

AGRO-ECOSYSTEM ANALYSIS

Agro-ecosystem analysis (AEA) is an early manifestation of the change in outlook that led to greater participation, yet still has much in common with its origins in formal ecological analysis. The latter emphasizes the need to identify key components and flows (mostly energy) within ecosystems, and some of the early and classical studies took place in the 1950s. Agro-ecosystem analysis is the extension of these ideas into agricultural-based ecosystems. The primary difference between the two is that in agro-ecosystems humans are the dominant and central component in the sense that human management and activity is by far the major influence. Clearly social and economic influences are factors that cannot be discounted, and the logical step is not to ignore them but to include them within the framework of the analysis. Such analysis arose in the 1970s as an approach to understanding the complexities, social as well as natural, of agro-ecosystems and the use of such insights to direct research and development programmes.

Two types of agro-ecosystem analysis are described by Conway (1986), and this dichotomy still prevails in much of the literature on participation:

1   agro-ecosystem analysis for defining a research programme;
2   agro-ecosystem analysis for development (see Figures 5.1 and 5.2).

Both analyses are similar and largely comprise an extractive process in the sense that they are geared to get information on the agro-ecosystem rather than necessarily do anything directly to modify it. Both involve six basic steps as set out by Conway (1986):

DEFINE OBJECTIVES

↓

SET SYSTEM BOUNDARIES AND DEFINE COMPONENTS

↓

COLLECT INFORMATION ON KEY COMPONENTS

Space   Maps, transects of areas covered by the system

Time   Variation in key components with time
          Physical factors (rain, sunlight)
          Cropping pattern
          Pests and diseases
          Prices
          Labour pattern

Flow   Patterns of transfer within the system
          Energy
          Money
          Information

Decisions Decision tree for farmers and others involved in the system
           Identify critical decision makers

KEY QUESTIONS OR GUIDELINES
(eg new varieties, input use, pricing policies and maintenance of infrastructure)

↓

IDEAS (HYPOTHESES)

↓

IMPLEMENTATION OF RESEARCH PROGRAMME
(feedback of results into steps above)

Source: Conway, 1986

Figure 5.1 *Agro-ecosystem Analysis for Determining a
Research Programme*

1   definition of objectives;
2   definition of system boundaries and components;
3   collection of information on key components;
4   key questions;
5   production of ideas, guidelines etc to provide solutions;
6   implementation/evaluation.

Information collection is by a combination of observation, surveys, interviews and use of existing data. Implementation is the last step in the process, and involves either the setting-up of a research programme or the establishment of a development project. Therefore, participation in this context appears to be little more than the involvement of farmers and others

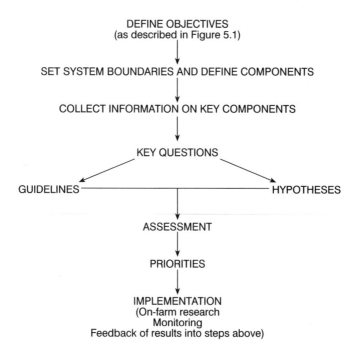

DEFINE OBJECTIVES
(as described in Figure 5.1)

SET SYSTEM BOUNDARIES AND DEFINE COMPONENTS

COLLECT INFORMATION ON KEY COMPONENTS

KEY QUESTIONS

GUIDELINES   HYPOTHESES

ASSESSMENT

PRIORITIES

IMPLEMENTATION
(On-farm research
Monitoring
Feedback of results into steps above)

Source: after Conway, 1986

Figure 5.2 *Agro-ecosystem Analysis for Development*

in providing answers to questions set by scientists. The agenda is still basically a researcher-driven one, with boundaries and key questions set by outsiders. However, the key point is that although AEA had many familiar elements for natural scientists schooled in traditional ideas of ecosystem analysis, it did begin to stress the need for farmer involvement, albeit in an extractive sense. AEA also began to stress the need for the inclusion of other professionals, particularly social scientists, as part of the research team.

FARMING SYSTEMS RESEARCH

Related to the rise of AEA was a broad move amongst natural and social scientists based in research stations towards involving farmers more in the research process. There are various names describing this process, reflecting surges in different locations and from different people, including Farming Systems Research (FSR), Farming Systems Research and Extension (FSRE), Adaptive Research (AR) or simply and more generally as On-farm Research (OFR). The basic ethos is that researchers work directly with farmers and incorporate their desires as far as possible. These on-farm dimensions may be a later component of an AEA as described above.

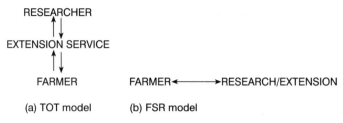

Figure 5.3 *Transfer of Technology (TOT) Model Compared with Farming Systems Research (FSR)*

In all these models the distinction between research and extension becomes blurred, and the key point is that the farmers play an integral part in the process. They are seen as partners with valuable knowledge as distinct from passive recipients of valued knowledge as in the TOT model (Figure 5.3). The form of OFR may vary in practice, and Biggs (1989) suggests that there are four modes depending upon the level and type of farmer involvement (Table 5.1).

Like AEA, these various farming systems research approaches tended to have a more extractive emphasis (ie the generation of new data – from the point of view of the scientists). However, in farming systems research there was also an ethos of working with farmers to address their concerns. In practice this typically meant testing a range of technologies under on-farm conditions with farmers providing their views on appropriateness along with suggestions as to how the technologies could be best adapted to local conditions. Typically the on-farm experiments were a compromise between the inclusion of farmer management and a perceived desirability for an element of experimental rigour so as to allow statistical analysis. Much of the literature discusses these trade-offs in terms of replication and precision. The collegial category described by Biggs (1989) was far less common. In that sense, the farming systems approaches had much in common with AEA in the sense of being top-down and primarily researcher-driven.

Table 5.1 *Four Modes of On-farm Research*

Mode	Objective	Contact with farmers
1 Contractual	Scientists contract with farmers to provide land or services (such as labour)	minimal
2 Consultative	Scientists consult farmers about their problems and then develop solutions	high at start and end of research
3 Collaborative	Scientists and farmers collaborate as partners in the research process	high and continuous
4 Collegial	Scientists work to strengthen farmers' informal research and development systems in rural areas	long term and sporadic

Source: after Biggs, 1989

RAPID AND PARTICIPATORY APPRAISAL

AEA and the various farming systems approaches were primarily driven by natural scientists and can be seen as extensions of research station science into the realm of the farmer. Familiar ideas and approaches to the scientist were modified and taken out to the farmers themselves.

In the 1970s another movement driven primarily by social scientists, began to initiate a dialogue with rural-based populations in order to better understand their circumstances and wishes. This broad process of rural appraisal began as rapid (RRA) and progressed to participatory (PRA). These two forms are quite distinct.

RRA is a term used to describe the practical steps needed to collect information on a defined system, and in many ways has much in common with AEA although it has a less ecological perspective (Walker et al, 1978; Conway, 1986). Another origin rests with a gradual disenchantment with formal social science research methods applied in developing countries. These tended to rely on standard tools such as questionnaires and surveys geared to generate data that could be analysed statistically. Hence there was a perceived need amongst social scientists to get good quality information more quickly. Yet another influence mentioned by some was a disenchantment with policies primarily implemented by economists. Other influences may well have been the application of community development within a revolutionary emphasis, most notably in South America (Freire, 1972), and writings on organizational structures and management.

Like AEA and farming systems approaches, RRA and its later offspring had an early emphasis on the rural population of developing countries, and particularly within agriculture. This broad concern over the link between agricultural research and development is often expressed as 'farmer first'. A conference held at the Institute of Development Studies (IDS) at the University of Sussex in 1978 set out many of the underlying principles and approaches of RRA (Chambers, 1983). Further elucidation of the principles of RRA and the 'farmer first' ideology can be found in the writings of Chambers (1992, 1993, 1997), Chambers and Ghildydal (1985), Chambers et al (1989) and Toulmin and Chambers (1990).

RRA is essentially an information-gathering process to feed back into policy, research and intervention. However, the intention is to attain this insight in a holistic sense rather than the reductionist approaches of so-called positivist science which were deemed to have generated the inappropriate research. In order to achieve this RRA is multidisciplinary, involving a team approach, and emphasizes both speed and flexibility. Methods vary depending upon circumstance, but the following are commonly applied:

- use of existing information;
- interviews (samples, key informants, groups);
- direct observation ('live' in the area);
- transects and mapping;
- diagrams (eg decision-making trees, calendars, timelines);
- ranking (wealth, income/expenditure etc);
- oral histories;
- surveys;
- activity profiles (eg farm labour profiles);
- workshops to help analyse/discuss issues raised.

By and large these are not far removed from the features of AEA and farming systems approaches. Unlike these, however, RRA was often portrayed by its proponents as being anti-positivist. As already mentioned, AEA and farming systems approaches were typically off-shoots of formal institutional-based research and were still arguably framed within a positivist agenda – albeit with more holistic ambitions. RRA was seen as different in the sense that it did not have a positivist agenda, although ironically the techniques of RRA are much the same as those used in the positivist studies they criticize! This oddity has been referred to by others, for example:

> *Thus the obvious limitations of RRA techniques – most notably their failure to depart significantly from conventional research on whose critique RRA was based – were explained away in terms of inappropriate social processes, that is lack of participation, and hence the move to PRA* (Sellamna, 1999).

Nevertheless, RRA became a popular approach, primarily because it can be applied to a diverse range of situations many of which are well outside the traditional emphasis on agriculture (Kumar, 1993), for example fuel wood, natural resource surveys, irrigation systems, primary health care and marketing. The specific mix of techniques employed depends upon the circumstance and the objective. Nevertheless, despite the emphasis on flexibility people have tried to catalogue what they perceive as different forms of RRA depending upon what it is trying to achieve. One example is provided by McCracken (1988), who suggests that there are six basic types of RRA (Table 5.2).

Although one can see the relative distinctiveness of some of these categories, at the same time there does appear to be a great deal of overlap. The first five types are primarily extractive in the same sense that AEA and the majority of farming systems approaches were applied. The sixth type illustrates the next step in the evolution of RRA. RRA began to be seen as having the potential to move beyond extraction and more towards

Table 5.2 *Six Types of RRA*

Type of RRA	Aim
1 Exploratory	To determine the essential nature of systems
2 Topical	To find answers to specific questions
3 Action	To enable an intervening action in a system
4 Emergency	Quick action after a disaster
5 Monitoring/evaluation	The assessment of impact
6 Participatory	To catalyse community awareness and action (same as PRA)

Note: Types of 'participation' are illustrated in Tabel 5.3
Source: after McCracken, 1988

community awareness and action. In other words the techniques could be employed to let communities learn about themselves and instigate change as a result. The new approach was termed PRA, and defined as:

> *a family of approaches, methods and behaviours that enable people to express and analyse the realities of their lives and conditions, to plan, themselves, what action to take, and to monitor and evaluate the results.* (IDS, 1996).

It is a mistake to think of PRA as somehow superior to RRA. They are quite different rather than representative of different levels on a evolutionary scale. Indeed, RRA as a means of gathering information may be a part of PRA. The difference between them is not in the tools they use but in what they are trying to achieve. RRA is an extractive process while PRA involves participative intervention and empowerment:

> *PRA is a means for actors to reach a consensual definition of what represents a 'problem' and what is an acceptable 'solution' and to jointly construct data in a transparent process* (Sellamna, 1999).

Alternatively, RRA and PRA may be seen as the ends of a spectrum from product (participation as an end – RRA) to process (participation as a means to an end – PRA). Hence techniques that depend on participation are central, and examples are participatory mapping along with shared presentations and analysis. The principles of PRA in practice stress the:

- unpredictability of social phenomena;
- the subjective nature of data; and
- the endemic nature of problems.

The first and last points are largely dealt with through a process of:

*   iteration;
*   no pre-set methodology;
*   flexibility; and
*   continuous and sustained work with communities.

Combined with these is the notion of 'optimal ignorance', or:

*   knowing only what is useful;
*   measuring only as precisely as necessary;
*   being appropriately imprecise.

> *For the complex, diverse, dynamic and unpredictable realities of people, farming systems and livelihoods, comparisons and judgement are often more potent and practical than precise measurement* (Chambers, 1997).

However, one should remember that the words useful, necessary and even imprecise are themselves value-laden and subjective. Who deems what is useful, necessary and precise, and how are disputed views negotiated? For example, in the case of participatory poverty assessment (PPA), Robb (1998) states:

> *PPAs should try and achieve 'optimal ignorance' and collect only that information which is of use to the policymaker.*

The subjective nature of data is dealt with via triangulation and this also accounts for multiple perspectives:

> *All views of activity or purpose are heavy with interpretation, bias and prejudice, and this implies that there are multiple possible descriptions of any real-world activity* (Pretty, 1998).

This similarity between RRA and PRA, approaches that their practitioners stress time and time again are quite different in outlook compared to positivist science, has lead to some interesting points:

> *The methodological proximity between supposedly different paradigms means that the same set of techniques can be indifferently used for 'conventional' data gathering, for participation and empowerment and for learning, depending on the profile, the wish, the will or the mood of the user.*

> *That is simply to rediscover an old truth: that methods are not to be reified and that they are only worth what their users are worth* (Sellamna, 1999).

This reaction towards a perceived dominance of scientific approaches may have some origin in a general disenchantment with science in the developed world. It is perhaps no coincidence that the anti-science stance within RRA/PRA arose at about the same time as concerns over the environment following various high-profile problems in the western world. Agriculture became agribusiness, and our food and the environment were seen to be but components in the process of making money. Scientists were seen as part of that process, and the increasing domination of biological and agricultural science by molecular biology has in turn helped to spawn yet another backlash. Yet despite this some have even asked whether RRA/PRA are really that far away from the very positivist and modernizing approaches that their practitioners decry. It is certainly true that RRA/PRA have features that distinguish them from positivist science:

- Approach: RRA/PRA is based more on induction as opposed to deduction;
- Reductionism: RRA/PRA is more holistic;
- Quantification: In contrast RRA/PRA is more qualitative;
- Institutionalization: RRA/PRA is by definition located in the places where the people are! It is the exact opposite of institutionalization.

Some of the most vigorous criticisms of positivist science by the PRA movement have been directed not at natural scientists but at neo-classical economists. Partly this may be a reflection of the power that these economists are perceived to possess, but is also a critique of their science for exactly the reasons given above:

> *Scientifically, economics is seen as the embodiment of an imperialistic methodology attempting to blend reality to a set of universal 'laws' patterned on classical physics.*
>
> *In economics, it is argued, concepts such as competition, market information, supply and demand etc, are presented as 'objective' facts, built in abstract mathematical models that act as a screen hiding power relations and value-laden judgements.*
>
> *These models protect economists from the messy nature of real-world data and represent a 'constructed reality' whose main features are reductionism and measurability* (Sellamna, 1999).

Three situations (1 to 3) with differing emphases on TOT and 'participation'

1   Mostly TOT with little or no client participation

2   Balance of TOT and client participation

3   Mostly client participation

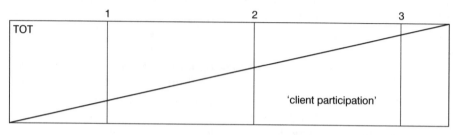

Source: after Garforth, 1997

Figure 5.4 *Blend of TOT and 'Participatory' Approaches*

Yet both RRA/PRA inevitably incorporate an element of simplification. Even if one begins with an holistic and inductive mindset along with a multi-disciplinary team, there is a limit as to how much can be learnt by all involved in a period of time. The term optimal ignorance may be a useful device, but does it not conveniently hide the inevitability of resource limitation driving what can be covered?

In practice a mixture of TOT and participatory approaches may be the best way forward as both have their own strengths. Garforth (1997) makes a convincing case for this and sums up the relationship in Figure 5.4. He goes on to suggest that that ultimate measure of success is what is achieved rather than how:

> *What matters in the end, I suggest, is that our actions – our efforts to institutionalize innovation(s) both as product and as process – should increase farm households' opportunities to achieve secure livelihoods for themselves by enlarging choice, enabling them to exert greater control over the factors that constrain their farming, encouraging them to experiment, and upholding the validity of local knowledge (Garforth, 1997).*

Whether the hard core of the participatory movement would agree with this is debatable.

CURRENT FRONTS OF PARTICIPATION

There has been much change within the participatory movement since the origins of RRA in the 1970s. The initial rural emphasis of RRA/PRA has since widened to include urban areas. While in the 1960s and 1970s poverty was a rural issue, in the 1990s it was much less so. Urban poverty will continue to be a growing problem in the 21st century, and there will inevitably be effects on priority setting with agricultural research (Trigo, 1997). The rapid element of RRA has also been disowned – some prefer relaxed instead! It is also interesting to note that both RRA and PRA have the word appraisal, and some have seen this as stressing their essentially methodological nature. Therefore a new term, Participatory Learning and Action (PLA), has been created to stress a move away from just rural and towards collaborative learning rather than just appraisal, with the learner:

> *'as an active goal setter, seeker and creator of knowledge, applying and building information into cognitive structures to solve problems in multiple settings'* (Sellamna, 1999).

Although the move away from just the rural is understandable, the dropping of appraisal has been seen to have more complex ramifications. Some suggest that it embodies a move away from an intervention agenda to a non-intervention and more philosophical one. Whether this is the case or not, some have used this perceived shift into the philosophical to paint a rather stark picture for the future of the participation movement (Sellamna, 1999).

Nevertheless, whatever the perceptions behind the move to PLA, there are examples that extol the more practical side of participation in ways that continue to reflect its strong interventionalist origins. The two specific examples discussed here are Participatory Monitoring and Evaluation (PM&E) and Participatory Poverty Assessment (PPA). Both are similar in that they build on, if not replace, traditional, well-established and non-participative approaches to monitoring and evaluation and poverty assessment respectively. A summary and review of PM&E can be found in Estrella and Gaventa (1997). PM&E dates back to the early days of the participatory movement in the 1970s, but has seen a steady increase in popularity throughout the 1980s and 1990s. In line with the participatory movement as a whole, the tendency in PM&E is to critique traditional methods of evaluation and monitoring as being positivist – based on some notion of objective and quantifiable truth. PM&E is seen as offering a different approach as, like the participatory movement in general, it aims to make the local people the primary focus. PPA is more recent, and was initiated by the World Bank as a means of allowing poor people, the real 'poverty experts' (Narayan, 1997; Robb, 1998), to express their views as part of a policy-framing process. The assumption being that: 'When their voice is heard, poverty alleviation efforts will have a much greater likelihood of producing the tangible results to which they aspire' (Narayan, 1997).

Many examples of PPA can be found in Holland and Blackburn (1998), and the results appear to be mixed (Robb, 1998; Norton, 1998). The tools and techniques for PM&E and PPA are much the same as those already discussed for RRA/PRA, and a central aim is to identify indicators (quantitative and qualitative) as 'a way of spotting and measuring underlying trends' (IDS, 1998).

A further and related advance worth noting is the rise of an approach referred to as sustainable rural livelihoods (SRL) or simply as sustainable livelihoods (SL) in the latter half of the 1990s (Carney, 1998). It has been known for long that the socio-economic dimension to sustainable development is important and one cannot just focus on narrow technical issues of natural resource management or environmental impact. The emphasis instead shifted to an analysis of livelihoods that encompassed agriculture and other non-agricultural activities, and these were analysed via a notion of assets. SRL has become very popular with aid agencies such as DFID. Within SRL, a livelihood 'comprises the capabilities, assets (including both material and social resources) and activities required for a means of living' (Carney, 1998) and, when sustainability is included:

> *A livelihood is sustainable when it can cope with and recover from stresses and shocks and maintain or enhance its capabilities and assets both now and in the future, while not undermining the natural resource base.*

The SRL approach makes much of the well-established fact that livelihoods are diverse: 'Studies show that between 30 and 50 per cent of rural household income in sub-Saharan Africa is typically derived from non-farm sources' (Ellis, 1998).

SRL begins with an analysis of the assets important to livelihood. Five principal ones are suggested (Figure 5.5): natural capital (land and water, for example); social capital; human capital; physical capital (infrastructure) and financial capital (savings and credits, for example). In SRL one looks at the vulnerability context in which these assets exist. In other words, what are the trends, shocks and stresses? After this there is the design of a suitable intervention linked to an assessment of outcomes. DFID favours the use of indicators as a means of assessing outcomes within the SRL approach. These are referred to as Objectively Verifiable Indicators (OVIs), and these are incorporated within the logical framework approach to provide a clear picture at the outset of what will be done, when and how outcomes are to be assessed. The indicators must be specific to outcomes, measurable, usable (ie practical), sensitive, available and cost-effective.

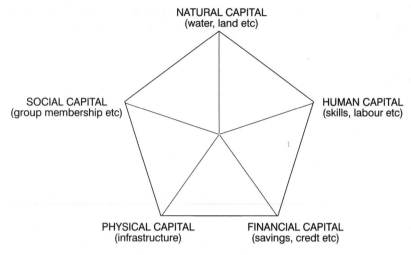

Source: Carney, 1998

Figure 5.5 *Assets Considered in the Sustainable Rural Livelihood Approach*

The use of indicators within broad and related spheres of measuring sustainable development, SRL, PM&E and PPA is increasingly being stressed (Bell and Morse, 1999; Morse et al, 2000).

SRL is essentially an approach to analysing and understanding the livelihood context of local people prior to planning an intervention. As such it is an extractive process rather than intended as a catalytic exercise in itself. In that sense it has much in common with the older AEA approach dating back to the 1970s. Even the language has a deep ring of familiarity. For example, sustainability is envisaged in terms of resilience to shocks – just as it is within AEA – and assets can be crudely seen as system components. The use of the SRL approach in a practical process of development project design and implementation is one of the reasons for its popularity.

THE FURTHER RISE OF PARTICIPATION?

Given the inequalities of power within the agricultural research environment of both developing and developed countries it was perhaps inevitable that the participation paradigm arose the way it did and became as popular as it has. There is also no doubt that it has some successes in the field of needs assessment for agricultural research:

> *Participation of farmers and extension agents in the analysis of socio-economic and production constraints and in the assessment and adaptation of new technologies has been shown*

to increase the rate of adoption of improved technologies by farmers (Monyo, 1997).

However, for all its appeal and successes there are some signs that the participatory movement has lost momentum. In particular, it is noticeable how thinking within the participatory movement has become crystallized in two areas: the renewed calls for the adoption of participation and the reinventing of labels (as in the move to PLA).

The first of these is an issue of use, and follows naturally from whether PRA/PLA actually succeeds:

> *While the practice of PRA has brought in a sometimes refreshing feeling of exhilaration (or panic) among the scientific community and forces it to reconsider its methods, the contribution of PRA, in itself, to the insights of research, the performance of development projects, or to the well being of the communities that use it remain still to be evaluated* (Sellamna, 1999).

As outlined above, participation is an element of many uses and not an end product in itself. AEA, farming systems and RRA were in essence extractive approaches designed to generate information that fed into the design of a development intervention. This may be, for example, the provision of information that could help formulate a research agenda or perhaps wider policy. PRA has similar leanings, although at the same time an aim may also be to facilitate internal change rather than inform outside mediated interventions. Inevitably there are concerns about whether participation mediated by an outside agency may just be a mirage resulting in part from a reluctance to provide the real picture. These concerns are magnified under typical constraints of time and resources. Also, attempting to heighten peoples' awareness of constraints may achieve little of value if opportunities to overcome them are also very limited. Bevan (2000) suggests that the 'first fundamental problem' with PRA arises because:

> *there is little recognition of the fact that poor people are diversely embedded in unequal meso, macro and global economic and social power structures, or of the fact that the passing of time entails trends, shocks and conflicts which lead to changes in structures, and in the positions of people within those structures.*

If people have no real power to change a situation, then highlighting the problems they face is not in itself going to change their circumstances. Even in a more limited information extraction sense, giving people a louder voice does not necessarily solve the problem if no one is willing or able to listen. This strikes at the very heart of participation:

> *Indeed for many who use PRA and related methodologies at the grassroots, getting the poor to participate has no meaning unless it simultaneously addresses the power structures that appear to perpetuate poverty'* (Blackburn and Holland, 1998b).

Yet, governments and aid agencies alike have multiple demands on their resources, and there are no guarantees that all demands can be met. Bottom up flows of information enter a maelstrom of institutional agendas and may not be the sole criteria upon which organizations act. In this maelstrom is it not regretfully inevitable that participation from some groups may have to be compromised? If it is compromised for some, then is participation incomplete? The result may well be that the poor become tired 'of being asked to participate in other people's projects on other people's terms' (Bevan, 2000). In its purest manifestation, participation thereby becomes a means towards an end of greater social justice rather than just a means of achieving better development as an end product (Pretty, 1998). Participation in this sense is a fundamental right – and this is an ideal that cannot be compromised. Pretty (1995) provides an interesting spectrum of the meaning of participation, with 'manipulative' at one end of the scale and interactive/self-mobilization at the other (Table 5.3).

This concern over the use of participation has been expressed many times before. For example, we have the following quotation from 1996 – nearly two decades after the drive for greater participation began:

> *The question is how to make the transition so that the insights coming from PRA begin to be translated into policies and practices that actually benefit the poor. This is no easy task given the entrenched attitudes and vested interests that are involved. To imagine otherwise would be naïve'* (IDS, 1996).

In other words, the question becomes how to institutionalize or scale-up participation so it is both carried out on a routine basis and can influence policy. This is not simply a matter of communication from bottom to top, but a recognition that there are many influences and pressures at play that have to be addressed (Blackburn and Holland, 1998a). Ironically, parallel concerns about the institutionalization of strategic planning in agricultural research have also been voiced about NARS in sub-Saharan Africa. Here the problems of inflexibility in planning have been highlighted – even when done in a largely top down mode with little or no participation of farmers:

Table 5.3 *Seven Types of People 'Participation' in Development*

Type of participation	Characteristic
1 Manipulative	A pretence (no real power). For example, the presence of 'people's' representatives on a board or committee, but who are outnumbered by external agents
2 Passive	People told about a decision or what has already happened, with no ability to change it
3 Consultative	People answer questions. The form of the questions and analysis of results is done by external agents, It equates to the 'consultative' form of OFR (Table 5.1)
4 Material incentive	People contribute resources (eg land, labour) in return for some incentive. The 'contractual' form of OFR (Table 5.1) is an example
5 Functional	Participation seen by external agents as a means to achieve goals (eg reduced costs) usually after major decisions have already been made
6 Interactive	People involved in analysis and development of action plans, for example. Participation is seen as a right and not just as a mechanical function. This equates to the 'collaborative' form of OFR (Table 5.1)
7 Self-mobilization	People mobilize themselves and initiate actions without the involvement of any external agency, although the latter can help with an enabling framework. This equates to the 'collegial' form of OFR (Table 5.1)

Source: after Pretty, 1995
Note: Linkages have also been made to similar categories in Table 5.1 (referring to on-farm research, OFR)

> *There remains the need to examine more closely why strategic planning in agricultural research fails to be institutionalised so that the product becomes the start of a learning process and not a static 'blueprint' for research'* (Hambly and Setshwaelo, 1997).

To date the prime focus with RRA/PRA has been very much on the rural people and what they want. In an agricultural context, the emphasis has been upon farmers and to a lesser extent upon methodologies of extension and facilitation through training, for example. A simplified research system is shown in Figure 5.5, where each of the boxes represents a complex set of interacting actors and agencies. Two-way flows of information are assumed, and clients (ie farmers) are on an equal footing with the formal research base and help to shape the research agenda.

Limiting factors are the feedback of information to the researchers and weak linkages between research and extension and extension and clients (Eponou, 1993). In summary:

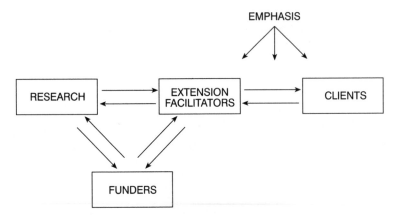

Figure 5.6 *Simplified Diagram of a Research System with Extension and Clients (ie Farmers) on an Equal Footing with the Research Base*

> *Identifying the agricultural technology needs of users, and translating those needs into research themes is an essential component of any priority-setting exercise. If properly done, a clear set of specific themes will emerge for ranking. If not properly undertaken, ambiguity about the set of alternatives to be compared is magnified in all subsequent steps of the priority-setting process'* (Audi and Mills, 1998).

Hence the main objective has been to create and spread a new mindset of participation and the tools and training to go with it. Understandably, most of the emphasis within the participatory movement has been on the right hand side of Figure 5.5, although not all constraints faced by farmers an be addressed effectively by agricultural research. Audi and Mills (1998) suggest that in practice the aim should be to address the overlap between the needs of the client and the set of problems that can be feasibly addressed by a research programme. There are various tools and methodologies that can be applied to determine this overlap, and one example is the use of 'problem trees' (Audi and Mills, 1998).

Even so, some have noted that despite this emphasis there have been gaps. For example, Okali et al (1994) suggest that there has not been enough appreciation of indigenous research systems within the broad participatory movement. Nevertheless, it has simply been assumed that the research base will respond accordingly once this information has been passed on. It does not take into account limitations and pressures that operate within the research base, and particularly how these interrelate with donors and wider strategic research planning. As highlighted by Hambly and Setshwaelo (1997) there have been relatively few studies that have attempted to analyse retrospectively the ways in which research plans have been set at national and regional levels in sub-Saharan Africa. What has been done suggests a great deal of variation between countries and generally, despite the rhetoric, little involvement of research users (Maredia et al, 1997). Given all of the emphasis on participation since the 1970s described in this chapter, this may be surprising and indeed disappointing. But as we have already seen in the context of the UK, there is a wealth of complex pressures and agendas that operate outside and within the research system, and some of these may act against a desire to increase participation: 'How to include stakeholders in research design and conduct remains a problem in developed and developing countries alike' Oehmke et al (1997).

The role of donors can be critical, and Mareida et al (1997) describe how some donors have attempted to influence the setting of research priorities by NARS to match their own views:

> *If research planning is primarily perceived as a donor requirement rather than a felt need of the countries themselves, the chances that the research planning process can result in a fundamental change of the research system's role vis-à-vis domestic research clients are much diminished. Research planning carried out in response to donor pressure may perpetuate the incentives for agricultural officials in African countries to consider donor agencies as the primary 'client' and respond to donor strategies for NARS development instead of investing time and effort in winning support from domestic clients and stakeholders.*

Given that external donors are an important source of research funding within the African NARS, the dangers inherent in seeing them as clients are very apparent and indeed have not been lost on the donor community. For example, a Sustainable Financing Initiative (SFI) has been launched by the Special Programme for African Agricultural Research (SPAAR), United States Agency for International Development (USAID) and others. The aim is to:

> *strengthen and diversify the financial base of African agricultural research and natural resource management institutions by promoting experimentation with new financial mechanisms and partnerships with clients* (World Bank, 2000c).

Although important, the influence of donors is not the only complicating factor. Figure 5.7 (based on Chambers, 1997) is often used to illustrate differing levels of status within an agricultural research chain that the farmer-first paradigm questions.

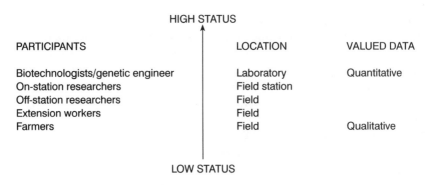

Source: after Chambers, 1997

Figure 5.7 *Hierarchy of Status*

Natural scientists, particularly those involved in more molecular fields of research, are typically based at the top (high prestige), while farmers occupy the base. Between these two extremes we have various levels of prestige, often related to the type of knowledge that is valued and the level of control of environmental conditions within which the research takes place. A particularly sensitive point in many countries is the ranking of extension beneath research, as it typically takes the form of substantial differences in salaries and incentives (Eponou, 1993). The aim of the original diagram presented by Chambers (1997) is to illustrate the problems that farmers face in such a prestige-driven research environment of making their voices heard, and there is, of course, much in this. Views such as that expressed by Pretty (1998) are typical:

> *Typically, normal professionals are single-disciplinary, work largely in ways remote from people, are insensitive to diversity of context, and are concerned with themselves generating and transferring technologies.*

As a result, and perhaps unsurprisingly: 'Their beliefs about people's conditions and priorities often differ from people's own views.'

However, it can be argued that given the wider context the entrenched hierarchy in Figure 5.7 and the characteristics summarized by Pretty (1998) exist for understandable reasons. Biology throughout the world has been moving into the more molecular sphere for decades, as exemplified by changes in UK research policy, and there are a number of reasons for this. Commercial interest in biotechnology, primarily in the health rather than agricultural arena, is one reason, and governments have been keen to improve linkages between the public-funded science base and companies. This is seen as a way of improving the wealth of the country and leads to job creation in a competitive world economy. Similarly governments have also been keen to obtain value for money from their scientists, and assessment (typically of output rather than impact) has increasingly become an accepted part of this. In a climate of accountability that increasingly permeates many of the cultures where taxpayers also want assurance that their money is being used effectively, this is not an unpopular move. Assessment has typically focused on published outputs for logical reasons, and, whether one agrees with it or not, journals at the more molecular end of the scale are typically seen as the more prestigious. It is no wonder that scientists seeking promotion or even to keep their jobs gravitate towards this end of the spectrum! While the situation is more complex than set out here, these two examples illustrate the logic that would inevitably lead to the ranking in Figure 5.7.

It is also unlikely that emphasis on assessment/accountability from those funding research will decline – the opposite may be far more likely. Dismissing the hierarchy in Figure 5.7 as entrenched and vested may be appealing rhetoric, but it fails to address the reasons why it exists. Any attempt to make the voice of the farmer louder in this chain may not necessarily remove these reasons. The result may well be frustration on the part of participation practitioners:

> *The implementation of a participatory approach provides no guarantee, in itself, about an organization's development priorities or how they will be met. This may seem uncontentious but is frequently under-recognised by some of PLA's proponents.*
>
> *The limitations of participatory methods become a problem where exaggerated confidence in their efficacy leads to their being used exclusively and uncritically* (Biggs and Smith, 1998).

> *The continuing use of physical targets to measure outcomes, together with the short time-frames that pressurize implementing agencies to 'produce' quickly, are shown to be incompatible with the process needs that are inherent to participatory projects* (Blackburn and Holland, 1998).

> *In the case of the PRA movement, there is a yawning gap between the scientific (and/or anti-scientific) elaboration's of the THEORISTS and the understanding, motivations and capacities of the bulk of the practitioners and donors* (Sellamna, 1999).

However, the situation is not hopeless, and the relative influence of participation in directing policy and indeed the means by which this input can be institutionalized is an area of much current debate (Holland and Blackburn, 1998; Blackburn and Holland, 1998a). There are many case studies where these have been attempted, but success has been mixed:

> *What the case studies show is that participatory approaches do not sit well with hierarchically structured organizations, and that institutional cultures are embedded in wider cultures that are not so easily transformed* (Bevan, 2000).

Continued calls for the adoption of participation are common throughout the literature. Often they are as a result of a perceived inertia on the part of scientists or policy makers, with no attempt to go further in analysing or explaining the observed inertia. This literature follows a similar pattern to that which one can see in the 1970s and early 1980s, and indeed in some of the chapters of this book. Participatory approaches are deemed to be good because of inappropriate research and failures in development, and this is followed by a critique of reductionist science that fails to take into account farmer knowledge. Lastly we have calls for more participation and for scientists to be more responsive.

PARTICIPATION, RESEARCH AND THE CGIAR

An interesting example of this debate is a recent discussion paper produced by the CGIAR (Becker, 2000). The initial commodity focus of the CGIAR described in Chapter 3 was seen by some working in the stations as very limiting, and given the rise of participation in the 1970s it was understandable that there would be various attempts at formalizing a role for participation in the research programmes. However, as Becker points out:

> *Some of these approaches were well known in several arenas, although, in the CGIAR, they were restricted to a few pockets. The mainstream of biological scientists within CGIAR remained highly skeptical and untouched.*

Since that time the CGIAR centres have been under an increasing donor pressure to institutionalize participation, but there have been problems. Becker (2000) lists seven:

1   The dominance of the technology transfer approach, such as T&V, in the 1970s and early-1980s;
2   The view that participatory approaches are a better way of doing technology transfer, and should be a function of NARS, extension and NGOs rather than CGIAR centres;
3   CGIAR scientists with real experience of participation are still a minority. Those that are in the system tend to be on short-term contracts, hence problems of continuity that actively mitigates against the 'long term' and 'continued contact' soul of participation;

4   A continuing narrow focus of CGIAR on the natural sciences, allied
     with the view already described that 'good science is natural science'.
     Social sciences are at best assigned a supportive rather than central role.
     This is not just a trend in the CGIAR (Maredia et al, 1997):

> With regard to human resources, the historical emphasis of
> agricultural research systems in sub-Saharan Africa has been to
> build staff capacity in the biological sciences. The social
> sciences, therefore, has seen a relative under-investment in
> human resources, compared to the biological disciplines
> (Mbabu et al, 1998).

5   Low level of commitment from senior management, largely as it views
     'participatory research as a donor fad and a misallocation of money';
6   Reward system in the CGIAR centres, as indeed it is elsewhere in the
     world, is very much based on the production of quality data that can be
     published in high-impact factor journals;
7   Prevailing commodity orientation of CGIAR centres that mitigates
     against a systems perspective.

Becker goes on to propose that each of these be addressed, although as
already discussed in Chapters 1 and 4, they are somewhat intractable given
the other pressures found in research systems. However Nickel (1997),
who has held senior positions in a number of the IARCs, describes what he
sees as the realistic future for the CGIAR stations:

> Centers are now forced to become much more aggressive and
> innovative in supporting their activities through sets of
> individual projects. While this has introduced greater
> discontinuity and demands much of the time of scientists and
> management, it has had the positive effect of creating greater
> accountability.

Clearly accountability is played off against exactly the sort of conditions,
such as continuity, that Becker feels are required for enhancing
participation. Nickel (1997), on the other hand, states that:

> As a greater proportion of funds come from countries that
> benefit from the research, funding will move towards the goal
> of being beneficiary-financed and demand-driven.

The reward system for CGIAR staff is an interesting example of conditions
that mitigate against participation and indeed the criteria for recruitment.

> *A difficult issue is the reward system of the CG [Consultative Group] as well as criteria for staff selection. There is little incentive for researchers to do participatory research. This is certainly not only a problem of the CG, but of scientific institutions in general. However, it seems that the CG is not at the forefront concerning a redefinition of what is considered to be successful research and a successful researcher* (Becker, 2000).

Yet problems surrounding the reward system and status of scientists within a participatory agenda have been raised by many others over a long period (see, for example, Chambers and Jiggins, 1986, Biggs, 1989, Cox et al, 1998; World Bank, 2000b) and a detailed report covering seven developing countries (including Nigeria) produced for ISNAR, one of the CGIAR centres, by Eponou in 1993 makes this point crystal clear:

> *A striking phenomenon in most of the technology systems studied is the flagrant incompatibility between professional reward systems and organization goals. While the raison d'etre of most of the publicly-funded research systems is to generate relevant technologies for farmers, their employees' reward systems do not reflect this. In some cases, researchers who devote their efforts to technology development or to establishing linkages with technology transfer may even be penalized.*

At the heart of the problem may well be an emphasis on indicators of research output instead of impact (Eponou, 1993; Biggs and Matsaert, 1998; World Bank, 2000b), with output often used as a proxy for impact. Even here the emphasis will be upon the standard forms of academic discourse (peer-reviewed journal papers and conference papers for example) rather than pamphlets or posters aimed at extension agents or farmers. This sounds perverse, and many including Eponou have called for a change:

> *Reward systems currently in use are not compatible with the mission and goals of technology systems. If performance of these systems is to be improved, top management should redefine the mission and goals, and design new reward systems that focus on technology generation and transfer.*
>
> *Sabbaticals, training and promotion must be based on the relevance and effective use of technologies, rather than on the number of scientific publications.*

Eponou (1993) looked at NARS within the countries concerned, but the irony is that in the seven intervening years between this report and Becker

(2000) there does not seem to have been any progress towards achieving a greater emphasis on input rather than output within the CGIAR system. Why should this be so? After all, there is no shortage of participatory frameworks and methodologies designed to encourage better linkage between researchers and those they are meant to include as equal partners – the farmers. Goodell (1984), Biggs and Matsaert (1998) and Biggs and Smith (1997, 1998) describe some examples geared towards natural resource issues.

All that would be required to compliment this activity is a shift in the attitudes of top management, yet even in the CGIAR with all of its resources this has not happened. For example, Becker calls for a continuation of donor pressure towards participation as a 'lever for change'. However, he goes on to say that 'It is important, however, that donor commitment to the issue has a long-term perspective with multi-year funding, if changes are to be substantial.' Is this realistic? Will donors provide such multi-year funding, or will they simply continue with the current dominant practice of short-term funding? A significant indicator would be a willingness to do away with the short-term (one- to three-year) contracts in research, but there have been such calls for many years and no progress has been made.

Despite all the rhetoric the use of output indicators such as number and quality of publications exists for very good reasons from the perspective of employers and research funders, and despite repeated calls (eg Eponou, 1993) it seems unlikely that this will change. It is rather difficult to see how the CGIAR could hope to be at the forefront of addressing these, given that it operates and recruits within an international market that strongly favours the very things that Becker and others regard as hindrances to greater participation. Even if the CGIAR centres were willing and able to remove these pressures within their own borders, scientists working there would not wish to make themselves unemployable to non-CGIAR institutions, and neither would they wish to diminish their perceived status amongst colleagues that work in such institutions. The scientific world is simply much bigger than the CGIAR, and while one can applaud any institution that tries to move away from these standard measures of performance, they will find it difficult.

THE LIMITS OF PARTICIPATION

The literature that involves a reinvention of participatory labels, often allied with a philosophical argument, is beginning to foster frustration even amongst adherents to the principle of participation. As Sellamna (1999) points out, the participatory movement is in danger of collapsing in on itself as we continue to see changes in labels describing much the same thing, with philosophical arguments that become increasingly bewildering. Rethinking the fundamentals of what participation means in an increasingly esoteric sense may be a useful mental exercise, but is hardly an aid to poverty reduction. Again, this could be argued to be partly a result of the academic reward system giving more merit to papers published upon the nuances within the language of participation, as opposed to practical or logistical problems in relation to project planning, implementation and delivery.

The participatory movement appears to have reached its zenith, and perhaps that should be no great surprise. After all, the major battles were won years ago when the basic ethos of consulting people was accepted by all. It then became a methodological issue and this in turn has fed other debates. SRL is but an attempt to move away from a narrow focus on agriculture to livelihood and eventually life with leisure included. They are all firmly located at the farmer–family–household scale, with the basic aim being to elucidate problems and constraints and identify potential solutions. In a sense the participatory movement is still fighting the relatively easy battles and looking for easy targets. The problem is now surely one of getting those with power to respond to these insights.

At one level donors do just that, with a general acceptance of the principle that poverty alleviation requires a targeting of funds to real constraints and problems from the intended recipients point of view. It would be a brave donor that says otherwise! Yet there is unlimited scope for interpretation here. Those who contribute financially may well influence donor research agendas. For example, no matter what the wishes of African farmers are some major UK donors have pulled out of funding GM research:

> *Europeans tell us it is too dangerous. They tell us: 'Africa, this is not for you. Keep off.' You in Europe are entitled to your own opinion. But I think it is dangerous when you tell everyone else what to do* (Wambugu, 2000).

Whatever the rights and wrongs articulated by the anti- and pro-GM lobbies, this debate has nothing to do with participation, or at least participation of the African farmers, but everything to do with politics, finance, media relations and image in the developed world.

Participation can only take you so far, and after that the 'needs assessment' dissipates and competes within a wider context. The new battles are those that allow a ring fencing of the results of participation in this jungle so as to prevent them from becoming diluted and lost, and that presents a far greater challenge than the participatory movement has had to face to date. Indeed, is it even possible? Are we not inevitably going to be faced with a trade-off and negotiation between all these conflicting demands? After all, there are many examples in this book where there has been little, if any, public participation in the setting of research agendas in the developed world – including Britain whose institutions have, ironically, been at the forefront of promoting such participation in the developing world since the 1970s. At the end of 2000 the Royal Society in Britain has announced a new programme, 'Science in Society', aimed at promoting more public accountability and transparency from science – not in Africa or Asia but in Britain itself! In view of all that has been said in this chapter, some quotations from the President of the Royal Society, Sir Aaron Klug, are sobering to say the least:

> *The dialogue is about science's 'license to practice'. Science is necessarily, run by scientists, but it is society which allows science to go ahead and we need to make sure that it goes on doing so.*
>
> *So we need input from non-experts to make sure we are aware of the boundaries to our license. And, conversely, we need good channels of communication if we want to extend those boundaries, for example into new areas of research, such as embryonic stem cells, or new research methods, such as genetically modified animals.*
>
> *The public has expectations of how it will benefit from its investment in research. We must be aware of these expectations, and we should pay attention to them when setting broad research priorities.*
>
> *Nevertheless, we must not lose our nerve and concentrate only on research that appears relevant. The research agenda should be set by scientists, and they should have the freedom to change it in the light of new discoveries* (Klug, 2000).

The fact that an institution as influential as the Royal Society feels moved to introduce such a dialogue at the beginning of the 21st century with an investment of £1 million illustrates the complexities involved in enhancing participation. Yet even here the flavour is very much one of scientists in charge with participation as but one, albeit important, input in setting agendas.

Finally, it may be wondered why it could not be possible to separate out agricultural research agendas that apply within a country and those that apply to developing countries? Ostensibly, and perhaps superficially, this is the case, and will be returned to in Chapter 6. In the UK, for example, DFID is the main funder of agricultural research in developing countries and its policies are quite distinct from the agencies and structures that fund research to be applied within the UK. DFID's research priorities, for example, are highly applied towards initiatives that will help to reduce poverty. At higher levels it is possible to create different policies, but it should be pointed out that many of the major funding agencies involved in sponsoring agricultural research are located in developed countries, and their aid programmes typically require involvement from their own nationals. These work within universities, research stations and other organizations that often have multiple mandates, sometimes with a prime institutional focus on development within the UK, and are not primarily and wholly devoted to research in developing countries. Personnel within these institutions inevitably have to juggle the differing demands of multiple donors, while at the same time responding to the pressures of accountability, for example that apply to their own institution. Separation at higher donor and policy level does not necessarily equate to such separation at the levels of implementation.

SUMMARY AND CONCLUSIONS

In this chapter we have seen how the notion of participation on the part of those intended to benefit from research arose as a result of an increasing concern amongst scientists and academics based in the developed world. Given the colonial experience with agricultural research, a separation between the majority of the farmers and the research agendas of the stations would perhaps be inevitable. Similarly, the post-colonial period and a continuation of station-based research would also inevitably maintain this separation – the Green Revolution perhaps being one exemplar. It was also inevitable that this dysfunction would be noted and addressed, and perhaps the breadth of different participatory approaches that emerged during the 1970s from a range of different groups is an illustration of this. This dysfunction occurred to many people of many professions in many countries, and the results were almost concurrent strands of approaches such as AEA, farming systems research RRA/PRA and all serving to increase the voice of the rural poor. This has resulted in the development of a diverse range of methods and approaches, often not far removed from those of positivist approaches criticized by the participatory movement. Amongst some of the specialized applications of participation have been monitoring and appraisal (PM&E) and poverty analysis (PPA). There has also been a movement to redefine the meaning of participation with a reinvention of terminologies.

While laudable in itself, participation may not solve many of the basic causes of underdevelopment. Listening is not the same as acting, although it is an excellent first step. It has become apparent that there are many other factors at play besides listening, and one need look no further than the trends that have occurred in the directions of the UK agricultural research policies to see them. Perhaps ironically, an increasing emphasis on participation from UK-based and public-funded aid agencies has corresponded with a decline in the participation of UK farmers in directing the public-funded research process. It may also be suggested that public participation in this process has also declined, or perhaps even worse to suggest it has never been high anyway. Hence the moves of the Royal Society in Britain towards introducing a 'Science in Society' programme at the end of the year 2000. Given that Britain has been at the forefront of promoting participation in the less-developed countries, and has also been highly involved in the generation of methodologies to facilitate this and associated debates, one could ask why this apparent schism in outlook has taken place? How can the same country come to promote the importance of public participation in some parts of the world, yet fail to achieve it at home in the same period of time?

This is not to say that there haven't been attempts to apply participatory methods and approaches in the UK. Indeed, perhaps the two best examples of attempts to apply lessons learned from the developing to the developed world are the introduction of microfinance schemes modelled on the Grameen Bank in Bangladesh, and PRA.

We would argue that the limitations of participation have much the same underlying causes throughout the world. At the heart is the reality that governments, donors and researchers based in both the developed and developing worlds have multiple pressures and agendas, and these often have to be traded so as to arrive at a balance that completely satisfies no one, but partially satisfies as many concerns as possible. The voice of the intended beneficiary, whether one likes it or not, is inevitably part of this pressure. This occurs throughout the developed and developing worlds for reasons that are both similar and to some extent understandable although that is not the same as saying that they are excusable. An emphasis on participation as though it only applies to the developing world may result in an incomplete analysis. The developed world also needs to look far more at itself in order to understand why participation has the limits it seems to have and how these may be circumvented. Ignoring them or assuming that they apply in some parts of the world and not others, could be counterproductive.

Chapter 6

A Compromised Participation?

INTRODUCTION

In writing this book we set out to explore the agricultural research process and its limitations within the context of agricultural development over the last century in both the UK and overseas. The broad aim has been to raise some of the issues surrounding the establishment, functioning and constraints of the researcher–farmer chain from the perspective of the agricultural scientist/researcher. Given that the nature and form of this relationship has changed with time, and has varied between countries, it has been necessary to explore the evolution of these linkages in something of a constrained context – the UK and its ex-empire – but much of what has been said would apply to other parts of Europe and elsewhere. Included within this temporal and spatial analysis has been a discussion of the external pressures that have acted as overall drivers, including their effectiveness and their limitations.

Chapter 1 set out the broad background to agricultural research – who does it and why – and asked a series of questions that illustrate the complexity of the issues involved; even with seemingly simple and straightforward questions such as these. Some of the more specific factors that affect and influence agricultural research were investigated – from the political environment through to what motivates individuals carrying out the work.

Chapter 2 reflected on the way in which agricultural research has developed in Europe, but particularly in the UK, its origins and evolution largely during the last century and a half, and how research has driven, and been driven by farming practices and the politics and social needs of the time. During much of its history, there was a strong link between the researchers and the farmers – the latter even established many of the research stations we see today.

Chapter 3, by way of contrast with Chapter 2, explored the origins of agricultural research in the colonies of the British Empire focusing on the African continent and particularly Nigeria. Here the drivers, and constraints, and perhaps in particular the beneficiaries were very different.

Initially the focus was on export crops, with the beneficiaries being the commercial plantations and companies. Food production for local needs was simply not a priority, although this did gradually change. In complete contrast to the UK itself, research in the colonies was very much a top-down process carried out in stations and farms that were physically and socially isolated from the vast majority of the farmers that comprised the bulk of the subsistence food production groups they were meant to serve. Independence did not markedly change this.

In Chapter 4 the shift in power in Europe from farmer to politicians and the development of agribusiness was discussed. The birth of environmentalism brought about by the widespread use of chemical inputs and other detrimental ecological impacts associated with agribusiness led many to question societies' apparent trust in science as the sole approach for researching agriculture. The response has, in the main, been aimed at attempting to reduce the detrimental impacts whilst maintaining yield gains. Furthermore, there has been a shift in focus towards genetics and more emphasis on issues surrounding quality, although it is debatable as to whose perception and criteria of quality is being pursued. The methodology and tools used, however, have changed very little and are still based essentially on the scientific method of enquiry.

Chapter 5 summarized the birth of participation in development. Since the late 1970s, the need for inclusion and participation of people throughout the researcher–farmer chain in the developing world began to be acknowledged and hence the rise of the participatory movement. The evolution of the movement was explored in some detail in this chapter. This movement has focused primarily on international agricultural research aimed at improving agricultural production in the developing world. However, even after 20 years there are problems in its implementation. Now, interestingly, participatory approaches are increasingly being highlighted in finding solutions for agricultural problems in Britain and Europe. Some even suggest that the furore over GM crops is in part a consequence of a breakdown in this participation.

In this concluding chapter some of the issues arising in the researcher–farmer chain which developed in the two different systems (ie Europe and ex-British colonies) are compared. We have discussed and illustrated what the main drivers were and how they changed over time. Subsequently, it is not so much a question of what is the mismatch between the processes in the two systems, but rather, what is it that allowed these drivers in particular to flourish whilst other potential avenues for development have been closed or ignored. When viewed in this manner, it becomes apparent that one of the central issues here is of power and perception and that it is those in the most powerful positions who have had and will continue to have their perceptions of the future of agricultural research realized. Everyone involved, from consumer to farmer, governments, researchers and NGOs for example has had varying amounts of power regarding the direction and implementation of agricultural research over time.

THE AGRICULTURAL PROCESS AND THOSE INVOLVED

Although there is no definitive point in history to begin an analysis of agricultural change, the repeal of the Corn Laws in 1846 is a perhaps a poignant place to start. As mentioned earlier, it was this political event, perhaps more than any other, which marked the beginning of free trade in agriculture and set the wheels in motion along the road to agricultural decline in Britain. This depression remained, despite minor ups and downs, until the reintroduction of a protective agricultural policy after the end of World War 2. The response of the agricultural industry was to seek solutions for their problems through the establishment of research centres and the application of science. Although this interpretation is undoubtedly true, it does not go far enough in explaining what was actually happening at the time. A brief look at the introduction of what is now one of the UK's most successful crops – sugar beet – tells a far more complex story.

By the late 1870s the relationship between farm workers, tenant farmers, landowners and the social structure of rural society was put under increasing strain as the economic impact of the depression took effect. It is not within the remit of a book of this length to give a fully comprehensive overview of the social impacts of the depression. However, it is safe to remark on the following: unemployment caused rural unrest and led to the formation of the first agricultural unions, tenant farmers could not pay their wages to the workers or their rents to their landlords and subsequently many farmers and landlords went out of business (Fletcher, 1961; Hutchinson and Owens, 1980; Newby, 1988; and Tracey, 1989). Whilst there was an increased emphasis placed upon research to find answers, this was not the only course of action being pursued.

Looking to their European counterparts it was realized that sugar beet production could provide an answer (Hughill, 1978). As a root crop, it was realised that sugar beet could alter the rotation. This was important for several reasons:

- Sugar beet could replace the turnip/mangel-wurzel with potato in the crop rotation;
- Sugar beet tops could be used as a feed for cattle as well as providing sugar for people;
- European experience had shown that cereals grown after beet produced a higher yield;

- A deep-rooted crop could improve soil structure in poorly drained land;
- Sugar beet would be a well needed cash crop for depression-hit farmers in Norfolk and East Anglia;
- The introduction of the crop would provide much needed jobs on the farm at times of the year when normally labour requirements would be minimal;
- The construction and running of factories in the countryside would also help alleviate rural unemployment and poverty;
- From a sociological point of view the new crop presented the possibility of several very attractive outcomes for the different classes in the rural community. Firstly, the labourers would enjoy fuller employment. Secondly, tenant farmers would be able to earn extra money to help them weather the depression. Finally, landowners would be able to secure their position in the rural hierarchy through closer involvement in the newly developed agricultural stations and factories – thus being seen as leaders in introducing new techniques to the farming system.

In effect this was a continuation of shifting agriculture into industry, a process that had begun with the wool and textiles industry in the north of England (Tracey, 1989). With factories in the countryside, stocks and shares and capitalization in the city, agriculture could attain a higher status rather than merely existing to provide cheap food for industrial urban labour.

However, due to lack of government support coupled with the problems of how to encourage farmers to grow a new crop and no one prepared to build processing plants, it was not surprising that early attempts to introduce the crop failed. It was not until 1925 and the introduction of the British Sugar (subsidy) Act that production finally got started in any meaningful way. The subsidy was to run for ten years (Hughill, 1978). The effects were remarkable. In 1923 there were 16,000 acres under beet. By 1930 there were nearly 350,000 acres grown by 40,400 farmers. In 1923 there were just two factories, but by 1930 there were 18. The number of workers in the factories increased to 9900 and an estimated 30,000 casual workers found employment in the sugar beet fields (MAFF, 1931). This subsidy was to become permanent in the form of the Sugar (Reorganization) Act of 1935 (Watts, 1970).

What this potted history tells us about agricultural innovation is that far from this introduction being related solely to any particular group, individual, methodology, technology or scientific research, it was in fact the result of a congruence of events including:

- Rich landowners prepared to invest time and money to seek solutions – some scientifically and some via observation of their European counterparts;

- Political action brought about by a desire to quell unemployment and social unrest. It was a Labour administration more attune to subsidies, rather than a Conservative free trade government;
- The effects of the World War 1 shipping blockade and the threat to food supplies. A growing realization that free trade allows cheap imports at the expense of being dependent on foreign government's policies;
- MAFF being partly staffed and influenced by landowners, thus in a position to exert their power;
- The ability to industrialize the crop;
- The agronomic advantages of the crop to the UK farming system;
- The assistance of a Dutch millionaire prepared to form a company and invest money to build the first modern factory in the UK (Hughill, 1978).

All in all, it appeared to be a win-win situation. The story does not end there, for a levy was introduced on the production of sugar beet for the establishment of specific scientific research stations to investigate all aspects of the production of sugar beet. Furthermore, the story of the development of this crop in the UK is possibly one of the best examples of the impacts of post-World War 2 industrialization of agriculture in the UK: the advantages of mechanization in reducing labour, ironically, one of the main reasons for introducing it in the first place. The development of monogerm seeds and the use of chemical inputs in the form of fertilizers, herbicides and pesticides, also reduced labour demands. Essentially, this story tells that research and innovation do not occur in a social vacuum and subsequently, any analysis of what agricultural research does and what it has achieved must therefore include the social context as one of the main drivers for its evolution.

Set within this supportive context the research process itself was well integrated both vertically (farmers to policy makers) and horizontally (between farmers and researchers), and was effected both from the top down and from grass roots up. Communication between different groupings was perhaps easier, because as stated above large farmers were often also politicians or businessmen, providing obvious lines of communication to government policy and industry. At the same time the politicians' closeness to farmers, or at least those that owned and managed the land, meant that indigenous knowledge was often also part of the solution. Landowners also belonged to major pressure groups such as the National Farmers Union and the Country Land Owners Association that at least outwardly had a commonality of purpose. Answers to problems were sought without reference to discipline, and generally the process was truly multi-disciplinary in nature – inventors were allowed to be sufficiently open with few strict structures confining creativity. Farmers themselves were much closer to the problem solving, and in fact they often were the inventors. In a way the process was inherently and truly participatory – for the farmers at least.

However, as the social context of the times changed so did research. During the 1950s with the increased use and development of petroleum-based agricultural chemicals coinciding with a time when society was supportive of the belief that science alone would be able to overcome agricultural constraints, a shift occurred in the relationship of farmer to researcher and within the research process itself. The result was an approach to problem solving more led by scientists rather than farmers or their representatives in a much more organized and structured way. Agricultural research became more discipline-oriented – in the 1950s the central role of DNA in inheritance was discovered – and in vogue science shifted not only from field level to laboratory level but also within the different scientific disciplines. During the late-1950s and early-1960s chemistry was king. As the environmental impacts of this approach and the problems of pesticide resistance grew a new direction emerged. Agricultural scientists became biologists seeking solutions within the fields of biological control and population biology. With the development of computers and the growing recognition of ecology as a discipline, research focus shifted towards systems analysis and integrated pest management – the use of all control methods including pesticides to reduce pests to an established predetermined level which would maintain yields and economic viability.

Consequently, the integrated nature of the agricultural research process with farmer involvement was becoming diminished – farmers began to lose their autonomy over agronomic decision making. With the aforementioned accession of the UK to the EU in 1973, and the associated problems of overproduction the push for increased food production was no longer a priority. Economically, a system of agricultural subsidy that provided the exchequer with an income before overproduction was rapidly turning into one that cost the tax power an increasing amount of money. Farmers having overcapitalized were caught on a treadmill – increased land prices and economies of scale require increased spending and thus borrowing on expensive machinery. Pressure grew on the research industry to come up with new approaches. However, as the research process had become discipline-orientated and top down, with diminished input from farmers, it was scientists primarily who were given the task of finding solutions. MAFF and the farming lobby, by now so used to assuming that science was the way forward, supported this shift.

With the applied research on agricultural systems beginning to go out of fashion, partially as a response to the growth of environmentalism and also due to government funding unwilling to support research that merely perpetuated the system in place, there was a move to more pure research on animal and plant molecular systems. Maybe genetics could hold the key to reducing the unwanted side effects of our production system whilst maintaining yields. By the end of the 1970s agriculture became agribusiness, and the processors and retailers became dominant forces. Politicians and agribusinessmen became spokesmen for farmers, but the lines of communication between them were much less clear than they had been. Gradually research began moving from the public to the private sector with an inevitable decline in openness. The importance of links to industry began to emerge as a major influencing force in the same way as it was during a much earlier era overseas.

Meanwhile, in overseas territories the agricultural research process was markedly different. Here the emphasis was on the production of cash crops for export and money generation. The research process was entirely top down and transplanted, with research stations set up by the British colonial government targeted at these crops. Government officials, plantation owners and commercial companies were the main players here, and – except for the plantations – farmer participation in any decision-making would have been minimal. Agricultural research was formal in style, and based on European systems, and indigenous knowledge was rarely included in problem solving; neither were farmers asked what targets for research they wanted. By way of contrast, these same farmers as a part of an informal research process implemented the vast bulk of research aimed at helping the resource-poor farmers. State-funded research was carried out by paid researchers in relative isolation, as opposed to the more informal scenario depicted above in the UK, where farmers were often the inventors and worked in close cooperation with station-based researchers. Even with a shift to food-crop research in the colonies, extension was viewed as a mechanism for the delivery of valued knowledge to farmers rather than an interactive process that allowed feedback and knowledge from farmers. There was a gradual shift of emphasis to subsistence farmers, but this brought many problems, particularly due to conflicts in approach. All of this became recognized in the 1960s and 1970s, and failure to achieve impact necessitated a move toward more participatory approaches.

The realization that large amounts of money being spent on development aid were not achieving donor's expectations initiated the search for methodologies that would improve development impact. A more holistic approach was seen as the key, which included the participation of those involved in the project, particularly the end users, in determining and setting the agricultural research agenda and targets to improve impact. Hence the beginning of the participatory movement that has taken many specific and evolving forms since its birth in the 1970s. Since the 1970s considerable work has gone into developing new approaches, including AEA, FSR, RRA, PRA and most recently livelihood analysis. The requirement to include all stakeholders in decision-making where the risk of project failure is high, makes much sense. After all, those intended to benefit from the help that the formal research base is intended to supply would be in the best position to judge what they want! Without a perception of individual ownership of a project, the motivation to ensure success is much reduced. It is understandable that all funders insist on it as a means of assuring impact in a sector so often doomed with past failure. Participatory methodologies have now become an accepted part of the agricultural research process in the developing world, and as such, they have also now become a discipline in themselves. One danger is that as participation is institutionalized it has a tendency to become inflexible and more formal, with the result being that participatory approaches come to resemble the very positivist approaches that their proponents denigrate – a total contradiction to the very basis for its inception.

THE ROLE OF AGRICULTURAL RESEARCH

The motivation for agricultural research obviously has much to do with the way it is done, and who is doing it, and this has varied considerably over the last century. As suggested above, research was much less organized at the beginning of the century; much less formal and certainly much less bureaucratic. In the UK the inventors – farmers or businessmen – needed to find a solution to overcome a particular problem, or to supply a particular demand. Similarly, although the agricultural process was markedly different, agricultural research in the British colonies was done in response to a need to find a solution to an existing constraint or to explore new export crop potentials. However, in both situations the potential gain from organizing research and putting more concerted effort into finding solutions and developing new technologies, became apparent. What had previously been done by many as a hobby or an interest was now recognized as having enormous potential economic benefit. Research began to be more structured, and research institutions were established along similar lines both in the UK and in its colonies. The national motivation was clear – in the UK there was a need to become self-sufficient in food production while in the colonies the motivation was entirely commercial.

However, over time and as knowledge expanded, scientific and agricultural research became a profession in itself, and the motivation shifted more towards the academic. On an individual level, scientists increasingly did research to improve knowledge about the natural world as part of a career. The basis for research was no longer entirely focused on finding a solution to a particular problem. It was now recognized that the most interesting technologies discovered were not done through entirely directed research, but rather through creative investigation. As stated by Sir Aaron Klug (2000), 'the microwave oven did not come from someone trying to make a stove more efficient, nor the laser from trying to make a brighter light'. Individuals were rewarded either by promotion or simply by keeping their jobs on the basis of their scientific knowledge and analysis rather than the usefulness of what they had generated per se. Application of this knowledge became less important. From a national perspective, the potential for wealth creation through creative discovery was at the fore, but at the same time an increase in agricultural productivity, meant that the need for increased food production disappeared, and government policy shifted away from agriculture to rural development and the environment. The necessity for agricultural research per se was less clear. Rather, the desired output became increasingly holistic in nature. This has caused problems of scale in trying to integrate an agricultural research process that spanned from research related to field-level biology and the environment at one extreme, to molecular biology at the other. The need for efficient and effective mechanisms for communication and feedback became paramount – a battle still being fought.

WHO ARE THE PERCEIVED BENEFICIARIES

So who were the beneficiaries in this, and have they changed? In the UK the obvious intended direct beneficiaries of agricultural research are supposedly the farmers, and at a simplistic level, by the turn of the 19th century in the UK, this would undoubtedly have been the case. They identified the problems, were involved in finding the solutions and would clearly have benefited from them.

From the 1950s onward, farmers progressively moved down the list of perceived beneficiaries. Increasingly it was the agribusinesses, the supermarkets, the consumers and the public in general who were seen as the end users of research. This trend has caused some severe problems, as in many cases these new beneficiaries are a highly diverse group with often very different agendas, have little knowledge or understanding of the agricultural research process and little participation in setting its goals. A notable example to illustrate the point is the recent controversy over GM food. Government and the GM industry repeatedly claim that the new technology will bring benefits to consumers, yet these were never consulted in setting research agenda. It was assumed by all those with the power to set these agendas that they knew what the consumers wanted – why bother to ask or even explain what is being done?

However, in the developing world farmers initially were by no means the intended beneficiaries of agricultural research with the obvious and intended limited exceptions of the export-crop plantations. As already stated, the beneficiaries of agricultural research overseas were undoubtedly those with commercial interests – whether the companies, individuals, or the colonial power. In the case of Nigeria it was undoubtedly the 'UK plc'. Later, (as early as the 1920s and 1930s in Nigeria), there was something of a shift to considering the local farmers, but an ignorance of production practices above all else still acted as the major constraint. Nevertheless, throughout the developing world there has been an increasing drive to more closely define the intended beneficiaries of research and bring them into the planning process. For the most part the intended beneficiaries have been producers rather than consumers. Even the more recent manifestations such as those based on the concept of livelihood and the assets that underpin it are not necessarily consumer-orientated.

Viewed thus, it would appear that as the UK has moved away from a research heavily dependant upon the direct input of farmers, whilst the opposite appears to have happened in the developing world with more and more emphasis on farmer participation. Why should farmer participation be good for setting research agendas in the developing world and not here. The answer lies in the power relationships and perceptions of those who are setting the agenda – ie, who funds it and why?

WHO FUNDS AGRICULTURAL RESEARCH?

The way in which research is financed, has clearly been the main driver determining the direction the agricultural research process takes. At the beginning of the century wealthy farmers in the UK would often have paid for the research required to provide the solution they needed. Agriculture was then a very important industry, even though it was primarily for food-crop production and had suffered somewhat from cheap imports. It was clearly worth investing money to provide a solution. Environmental concerns, as we know them today, did not have a prominent role, but no doubt individual farmers and landowners saw themselves as custodians of their land.

As the UK became self-sufficient in food production, the shift in policy was away from agricultural research in general, but in particular away from improved yield towards environmental issues, supermarket quality demands and means of cutting costs. Farms became fewer and larger, and the number of people directly involved in the farming business decreased. Farming was only profitable if the farm was of a certain size to show economies of scale. Again, the need for a balance between agricultural production and protection of the rural environment became clear. With this shift, the power of consumers and supermarkets began to show as producers became relatively small players in comparison to the food industry taken as a whole. The supermarkets became a very powerful economic force – for example in the year 2000 Tesco had ten times the income of Monsanto, one of the largest agribusinesses. The power shift was clear.

The institutionalization of the agricultural process in the UK and the means by which it has been assessed has had a major impact on the way in which the research system now operates. Methods of assessment have become increasingly focused on individualistic performance criteria such as the number and quality of scientific publications, measures of international esteem with regard to scientific excellence and novelty of research as judged by peers, for example. Impact, technology transfer, and uptake of the technologies developed were rarely assessed, perhaps partly because they could not be so easily linked to an individual researcher. This is in marked contrast to the stated aims of the agricultural research process currently operating overseas through international development funders where lines of communication, participation by end users and stakeholders, technology transfer, uptake, development impact and sustainability are becoming of overriding importance.

It could be argued that the same assessments are needed to be able to judge wealth creation in the UK. However, although the stated aims in developing countries were more impact driven, in practice the same esteem-based criteria as those used in developed countries such as the UK became the norm. Perhaps this should not be surprising given that professional scientists have to work in an increasingly global system of research. Ironically the need for an individualistic assessment of researchers may have been partly driven by the increasing move towards short-term contracts as a means of maximizing flexibility. Funders understandably want results in a defined timescale for a defined cost. Increasing the mobility of scientists may be good at one level, but this will mean that they will want to make themselves marketable in this global arena. If esteem-rather than uptake-based criteria are central to this then it is no surprise that the latter has been neglected – no matter what the rhetoric.

Overseas, the prime motivating forces were initially the colonial governments and private enterprise (plantation owners and associated companies), whose objective was the creation of wealth through exportation. The impact of industries and companies was considerable – it was a very hierarchical and extractive system based on laissez-faire economics. After World War 2, the system changed with increasing emphasis on centralized approaches, particularly in the colonies. Laissez-faire was no longer seen as a panacea. In the 1950s the first agricultural research council for the colonies was established, and soon afterwards the colonies throughout Africa and elsewhere began to attain independence. The research process continued to be government-led, but this time, it was self-government. Independence, however, should not be confused with thinking that these new states were left alone to determine agricultural policies that represented the views and needs of their electorates as opposed to being outsider-driven. With the onset of the Cold War, agricultural policy had become a part of foreign policy for both the West and the USSR. The political importance of the ability to persuade or influence countries' political determination through preferential trade agreements and the establishment of research centres to help countries provide exports whilst increasing domestic food supplies must not be overlooked. During this period there was a global change from a system of international free trade to very controlled trade as established under the Common Agricultural Policy. Following the signing of GATT and the establishment of the WTO this began to shift back towards free trade.

Similarly, by the 1960s, with growing international concern about famine and long-term global food security leading to the establishment of multiagency funded regional international research centres as part of the CGIAR, focusing on subsistence food crops and systems, it should also be borne in mind that the global politics of the time as much as, if not more so, than any sense of altruism was probably the major influence. This may be being a bit harsh on the genuine concerns of the UK and Western society and the actual researchers working for solutions to food shortages. However, this is probably not a harsh reflection upon the perceptions of the then world leaders. It is also worth pointing out here that explanations for the causes of famine have also moved on from the simplistic idea that increasing crop production will end famine. It is now well accepted that a famine is as much a social phenomenon as a natural one, with the works of authors such as Sen (1981) and his theories relating to the entitlement and access to food providing a much more comprehensive explanation.

With the recognition of the problems of institutional structures, bureaucracy, and political agendas overseas in development projects, another group of players evolved. These are the NGOs who originally started out working at a grass-roots level, in a more direct way with less rigid structures. They gained strength as a lobbying force because of this. Recognition of the need for participation at all levels, and a way of working around immovable research systems led to NGOs being used increasingly as a vehicle for working outside these structures. Though it can be argued that whilst the work they carry out is well intentioned, it does to some extent allow national and international agencies and governments to reduce their commitment and evade their social responsibility in relation to resolving agricultural development issues.

ACCOUNTABILITY

The structures and systems arising to direct, monitor, assess and judge agricultural research have had a clear influence on the research process that develops in different situations. Accountability is an important aspect.

In earlier discussions, systems and procedures set up to monitor accountability have been shown to be key in determining the resulting research process. In the UK at the beginning of the century, there was a requirement for direct application of knowledge and problem solving.

There were very few formal structures in place, and this meant that accountability was in terms of achieving a successful solution to a practical problem, or in developing a technology to fill a market niche. Initially this was probably also the government's intention for organizations set up to do agricultural research, and systems were set up to measure accountability accordingly. However, science has become more and more discipline-oriented, and measures of accountability focused more on the science and the scientist. There is little assessment in terms of application of knowledge or even as impact in terms of wealth creation, national or international development. Rather, as we have seen, success is viewed in terms of publications, and international scientific recognition. Impact is judged in more narrow terms of the number of times a paper is read or quoted, rather than on the impact that the knowledge has in terms of agricultural production, rural development or agribusiness. In a world of agricultural research that has become very complex and broad ranging, perhaps it is too difficult, or inconvenient to set up procedures for accountability other than in terms of the science itself.

Perhaps it is in part because of this, that the general public has recently become such an important force in terms of measuring accountability of research in terms of application. The concern by the public to protect the environment has played a major role in accountability. In the 1960s following the problems with use of organo-chlorine pesticides, and the publication of Rachel Carson's *Silent Spring* in 1962, the potential to develop technological solutions that have a detrimental impact on the environment was widely publicized. Issues of global warming, climate change, destruction of rain forests are all emotive issues. During the last decade alone, the role of the public in putting pressure on the research process, firstly in the case of the BSE crisis and then regarding the development of genetically modified organisms (GMOs), is a familiar one. In both cases, procedures and structures in place to measure accountability were insufficient to satisfy a concerned public. The need to take these issues into consideration when commissioning agricultural research was clear and this is now a major influence on agricultural research. In a way this might be considered a form of participation. However, a concern frequently voiced is that this is not genuine participation – rather, according to Jules Pretty's classification of participation, that it is passive (people told about a decision or what has already happened, with no ability to change it). Furthermore, this form of participation is to a large extent media-led and therefore the information presented is often of a biased nature. In addition all too often the consultations seem to be hijacked by certain groups that steer or manipulate discussions to achieve very different political goals.

Internationally, development aid is big business and accountability is a big issue. The sums of money involved are enormous, and rightly or wrongly, over the last decade there has been considerable negative publicity with respect to the use and abuse of development aid, particularly when it is linked with trade agreements. This has fuelled the lack of confidence in the development sector in general, and probably as a result, increased the pressure to establish mechanisms to monitor and measure development impact – to do something that addresses the real images of distress that fills the TV screens and printed media of the developed world. Accountability has become increasingly important. People want to know that the money they have provided, either directly or through taxation, has had a real impact and not been wasted on bureaucracy. NGOs often vie with each other to show who has spent the least proportion of income on administration. Ironically, however, the need for greater accountability often results in more bureaucracy! With regard to agricultural research, participatory methodologies, and emphasis on effective communication in the researcher–farmer chain, have been increasingly supported as a vehicle to achieving greater development impact and ensuring accountability. Methods for accountability in terms of impact on poverty alleviation are, however, difficult to gauge, and the subject of considerable donor discussion. Yet it can be argued, as in Chapter 5, that although this is the rhetoric one hears, in practice it has been difficult to move away from the same prestige indicators used in the UK and elsewhere. Although it is broadly true that development-orientated research has more of an impact focus, it has proved to be difficult to move the research base as far down this road as donors may like.

PUBLIC PERCEPTIONS

The way in which agricultural research is perceived by the public has already been referred to. It has been, and continues to be, quite dynamic. The change in attitude since the end of World War 2 is a good example of this. In the 1960s and 1970s there was substantial global support for initiatives in international development, both politically and from the general public. Global political stability was seen as paramount, and improved agricultural production was seen as key to achieving this, by overcoming global food security problems and hence world poverty. Agricultural research and development was accepted as playing a major role in this. It was in this era of confidence that the CGIAR was established and the advances of the Green Revolution in Asia occurred. There are now many IARCs throughout the world, and they have an impressive reputation. However, global confidence in supporting this type of work has been lost to a great extent. There is a lack of confidence in appropriate distribution of aid, overkill by the media on emotive reporting, lack of good evidence of success stories, a move by rich nations towards a more inward looking perspective, the ending of the Cold War and the subsequent move to global capitalism and free trade. This has had considerable impact on support for international development in general, and the need for improved agricultural production.

In European agriculture the power of public perception has been made very evident through discussions on GMOs. Genetic modification is probably the most significant new agricultural technology in the last two decades. Its potential impact in altering the face of agriculture is enormous. Considerable sums of private and public money have been invested in its development over decades. However, public concern that this is an inappropriate technology in terms of potential risks to human health and the environment, as well as concern for ethical reasons, has slowed the development of this technology. The impact of this public voice has been substantial:

- It has forced an unofficial EU moratorium on commercialization of GMOs; public referenda have been held in at least two European countries with regard to the technology;
- International export and import of major crops (maize, soybean) have been determined by it;
- Numerous small businesses have gone out of business because of it;
- Supermarkets have reassessed their supply chains on the basis of it.

SCIENCE AND SOCIETY

Perhaps it has been the scale of the debates regarding GMOs and BSE in the UK, and Europe in general, and the realization of the power of the people that has catalyzed governments and funding bodies, as well as the scientists themselves, to rethink their interaction with the public. A large part of this has come in terms of a view that there is a need to educate the public in what scientists are doing and to explain the motivations behind the technology development. However, increasingly there is also an awareness that the way in which research is prioritized and planned is also critical, and that perhaps the general public has a role here.

In 2000, the House of Lords Science and Technology Select Committee produced a report 'Science and Society'. This explored the topic of the role of the general public in scientific controversy, and stressed that the dialogue is about science's license to practise. As stated in the President of the Royal Society's Anniversary Address (Klug, 2000):

> *It is ultimately society that allows science to go ahead and we need to make sure it goes on doing so... we need input from non-experts to make sure we are aware of the boundaries to our licence ... and we need good channels of communication if*

> *we want to extend those boundaries ... to new areas of research... Moreover, the public has expectations of how it will benefit from its investment in research. We must be aware of these expectations, and we should pay attention to them when setting broad research priorities... What we need to do is to engage in consultation and dialogue... The Royal Society's new initiative responds to a demand for more public involvement in the uses of science and more public accountability and transparency from science. It is part of science's social contract.*

At a European level, discussions on future Framework Programmes began following the publication of the document 'Towards a European Research Area' (European Commission, 2000). This explored the rationale for doing research in a coordinated manner at a European level, possible weaknesses in the past, and challenges to address over the subsequent four to five years. Interestingly, one of the sections focused on tackling the questions of 'Science and Society'. It suggested that:

> *More consistency should be introduced into foresight excercises, science and technology research, socio-economic intelligence and scientific and technological solutions taken at national and European level... there is a need to establish a platform of exchange, to create points of synthesis, and to align methodologies... The development of new and sustained dialogue between researchers and other social operators should also be encouraged... On the initiative of the national parliaments, in particular, measures designed to open a direct dialogue between citizens, researchers, experts, industrial managers and political decision-makers have been initiated... these formulas have illustrated the capacity of ordinary citizens to express valid opinions on complex issues and the possibility for groups with divergent interest to reach a concensus... Exchanges of experience that have taken place in this arena should be encouraged... Cross-participation formulas should also be tried...*

This is a clear call to bring genuine participation into the research process at a European level. Following national responses, a subsequent document 'Making a Reality of the European Research Area: Guidelines for EU Research Activities' was published by the EC. In this, the call for 'participation' was reinforced. Under the theme 'support for policymaking and European scientific reference system' it highlights: 'initiatives to involve stakeholders'. 'Further, activities are highlighted under the 'Science/Society dialogue'. 'Initiatives to bring into contact researchers, industry, policymakers and citizens...initiatives to promote the public's knowledge of science and technology'.

Is the intention for this process to be interactive? Is it acknowledged that the researchers have much to gain by listening, as well as explaining? The answers are unclear. The rhetoric suggests more of an enlightenment process than genuine participation.

At a global level, and focusing now on agricultural research in the developing world, the value of participation is increasingly recognized, in a manner that involves all stakeholders, including politicians. Here at least there is little contradiction in intent, although the means and extent by which participation is achieved, its institutionalization and its effectiveness in facilitating better development are all areas of much interest and activity and critical discussion. In 1996, the Global Forum on Agricultural Research (GFAR) was initiated by the CGIAR to mobilize the world scientific community in its effort to alleviate poverty, to increase food security and to promote the sustainable use of natural resources. Representatives include international agricultural research organizations; Advanced Research Institutes' (ARIs); NARS; donors; the private sector; NGOs and farmers' organizations. Its goals are to:

- facilitate the exchange of information and knowledge;
- foster cost-effective, collaborative partnerships among the stakeholders of agricultural research and sustainable development;
- promote the integration of NARS and enhance their capacity to produce and transfer technology that responds to users' needs;
- facilitate the participation of all stakeholders in formulating a truly global framework for development-oriented agricultural research;
- increase awareness among policy-makers and donors of the need for long-term commitment to, and investment in, agricultural research.

In response to this the European Forum was established in 1999 to strengthen the response of European agricultural research to development needs to ensure that the European voice was not overlooked. Its aims are, similarly to:

- promote European coordination, complementarity and comparative advantage;
- encourage participation of all stakeholders;
- assist the exchange of information and knowledge;
- enhance awareness among decision-makers and the general public.

The Global Forum and the European Forum were set up to facilitate participation by stakeholders at the European and global levels. It is a form of participation that differs somewhat from what is currently being advocated in UK and European research. Here it is the need to engage the general public in the research process that is the priority. This emphasizes the different backgrounds that are driving these moves. The drive in the UK and Europe for participatory methods is greatly influenced by the impact of the GMO and BSE debates on the scientific community and the failure to realize the importance of involving the public throughout the research process. In the developing world, the drive to involve all stakeholders in the research process has been the need to achieve impact in the context of international development.

THE POLITICAL AGENDA OF RESEARCH FUNDERS

The one overriding factor in all situations, is obviously the need for funding to support the research process, whether this is coming from governments, industry or development aid. It is the way in which external forces influence the funding agencies, and set their agendas that has the greatest impact on the research process. Funding for research overseas has always been rather entangled in international political agendas – from the situation during colonial times, when the aim was clearly to make as much money for the UK-based individuals or companies as possible, until the current day when development aid is only one part of a series of complicated foreign policies. As stated in DFID's White Paper (2000) 'too much global development assistance is used to sweeten commercial contracts or serve short-term political interests'.

For example, in 1998 large countries in South Asia received only US$10 of development assistance per poor person per year, compared to US$905 in the Middle East and North Africa (World Bank, 2000b). The requirement for accountability means that UK/European governments must be able to justify allocation of funding to development, in the context of benefit to the UK for the long-term financially or politically. Justification for governmental altruism is difficult!

During the Cold War it was easy to see the way in which donor allocation of development aid was influenced by international politics. Projects were funded in countries of strategic military importance rather than countries where the need was the greatest in terms of poverty alleviation. Agendas were not so obviously hidden. For example, the first Green Revolution in Asia during the mid-1960s, was allied with a shift in US policy away from using its own production to 'stopgap' potentially disruptive food shortages in the developing world and towards the encouragement of agricultural development within those countries (Cleaver, 1972). There was a search for political stability, and a need to maintain a degree of western influence on large, and potentially very powerful, countries. The need for global food security was viewed as paramount, and as such, there was much talk of eliminating hunger and poverty at a stroke (Cleaver, 1972).

Country loans with strict conditions are an example of the enormous impact of international and national politics on donors' agendas. Local people such as farmers do not set the conditions for the loans, and their participation is minimal. This is evident in the move by richer countries towards open markets and free trade – the increasing importance of the WTO has placed considerable pressure on poorer countries to follow suit. Country loans from international development agencies have been instrumental in taking this forward, and the SAPs initiated by the World Bank and International Monetary Fund are good examples. For example, in the mid- to late-1980s Nigeria was provided with large multi-donor loans, which came with the conditions of a decrease in government spending on subsidies, a reduction in government agencies and parastatals and the flotation of the Naira. Removal of expensive fertilizer subsidies was stipulated as one of the requirements. Farmers and researchers disagreed with the political rationale, and maintained that in a country with low soil fertility, the population would only be able to feed itself if fertilizers were used. As fertilizers were too expensive for farmers to buy at world market prices, subsidies needed to be maintained in some way. Removal of subsidies and the decline in value of the Naira had an enormous impact on the poorest of the poor, and only the rich could afford to use them. A similar situation has also been seen in Malawi. Exerting political influence by allocating funding according to criteria such as good governance, poverty focus is currently a more acceptable way of controlling agendas. However, this heavy-handed top-down approach to allocation of funding has often been less than helpful in trying to achieve meaningful development results. In trying to have the greatest impact on improved agricultural production, the researcher–farmer chain needs to be extended to the politician formulating the policies. It is worth noting that some of the lessons to be learnt by looking at past policies such as these, are now being translated into current policy statements. For example, in DFID's White Paper, it is clearly acknowledged that over-prescriptive aid conditionality has a poor track record in persuading governments to reform their policies.

Traditionally international project development has tended to be very top-down and the agendas of international agencies have had a massive influence. The focus of the largest funding agencies such as the World Bank, EU and FAO who support development aid in general is international development and the creation of national wealth and prosperity in a global context. Their policies are determined primarily by the international development policies, and these are obviously dynamic. Currently, targets set by the United Nations to reduce by one half, the proportion of people living in extreme poverty by 2015, is having enormous influence on the donor community. In the UK, DFID has framed its entire development strategy around these targets and the eradication of world poverty. It is hard to argue against such targets, even if they are overly ambitious. In order to achieve them, it is clear that the poorest of the poor are the main targets for development funding, and projects which have an immediate impact on poverty statistics related to basic health care and education are top of the agenda. However, a concern beginning to emerge is that this short-term focus is becoming all encompassing, and longer-term solutions that are critical to the sustainability of international development, are suffering. Research that would provide appropriate solutions for problems of agricultural production, is thought to provide solutions that are too distant in order to address immediate poverty issues. There is a school of thought suggesting that agricultural production is not a development priority – rather that sufficient knowledge is already available for appropriate levels of agricultural production to follow automatically. The requirement for those in the researcher–farmer chain to justify their existence in terms of poverty impact in such an environment, acts as a very powerful driver. The need for clearly demonstrable effective links between researcher and farmer that have an impact on poverty, is clear.

The donor emphasis on poverty alleviation is highlighted through DFID's policies regarding 'Sustainable Livelihoods'. DFID's 1997 White Paper on International Development committed it to supporting 'policies and actions which promote sustainable livelihoods' and is one objective designed to help achieve the overall aim of poverty elimination. Since 1997, various groups within DFID have been working to develop a better understanding of how to effect the sustainable livelihoods objective – and this has involved extensive consultation with partners as well as reflection on early efforts to implement 'sustainable livelihood approaches'. Sustainable livelihoods is a way of thinking about the objectives, scope and priorities for development in order to enhance progress in poverty elimination. It is a holistic approach that tries to capture and provide a means of understanding the vital causes and dimensions of poverty without collapsing the focus onto just a few factors such as economic issues or food security. It also tries to sketch out the relationships between different aspects of poverty, allowing for more effective operations (DFID, 1999).

In DFID's most recent White Paper (DFID, 2000) there is an interesting shift of policy. The emphasis on poverty elimination is still there, but there is a move towards harnessing globalization to bring benefits to the poor through private sources, provided there is support for capacity building for exploiting technological advances, participating in markets and establishing regulatory procedures. This will supposedly enable economic growth and thereby provide the potential for poverty reduction. The increased emphasis on the importance of commerce, trade, and private sector investments and partnerships is notable. The newly industrializing Asian countries such as China, who 'seized the opportunity offered by more open world markets to build strong export sectors and attract inward investment' are held up as exemplary.

Encouragingly, the importance of R&D in development is highlighted:

> *Most research and development capacity is in developed countries and is oriented to their needs. Research that benefits the poor is an example of a global public good which is underfunded.*

And interestingly, the need for private–public partnerships is also acknowledged:

> *More and more research is done in the private sector, and the low purchasing power of poor people means that there is little commercial incentive to invest in research to meet their needs. Governments and development agencies must therefore work to create more partnerships and must also invest directly and substantially in research that benefits poor people.*

With regard to agriculture:

> *… because most enterprises are too small to do research, publicly funded research remains important. The work of the Consultative Group on International Agriculture is vital. It is essential it moves forward with reforms to its government, organization and structure… Efforts must be made to strengthen the capability of developing countries to produce, adapt and use knowledge, whether produced locally or internationally.*

There is considerable discussion on how the poorer countries of the world can be brought into the global market place through responsible macro-economic and financial policies, through 'consultations beyond the Organisation for Economic Co-operation and Development (OECD) with developing countries on taxation issues…' and by promoting 'corporate social responsibility, particularly with regard to social, environmental and ethical policies'.

The approaches by DFID recognize the need for involvement of poor people themselves in identifying problems and finding solutions. It emphasizes the participation of stakeholders in general, at local, national and global levels. DFID's Renewable Natural Resources Knowledge Strategy research programmes, which commission agricultural research on behalf of DFID, are in the process of engaging with the sustainable livelihoods approach, including the conceptual framework developed by

DFID. This reflects a growing awareness among concerned research programme managers and advisors of the need for research programmes to demonstrate their potential contribution to poverty eradication, and DFID is currently supporting the further development and application of the sustainable livelihood approach within its programmes. To date activities in this area have focused on concept development, methodology development, and documentation of project experiences based on experience of using this approach. It is a good example of the way in which donors can influence participatory project development, through the provision of appropriate tools. As discussed in Chapter 5, without donors taking the lead to ensure that the poor are the prime clients, project focus can be misdirected.

In the UK, research agendas are set largely by government and European policies. Those who set these policies obviously have an enormous influence on the research that is subsequently funded. European agricultural policies have been established through the CAP. The question of who actually influences the government is not clear, although an increasing influence brought about by corporate lobbying is undoubtedly a concern here.

There is a lack of formal coordination between the various arms of the UK government charged with commissioning research. DFID commissions research as part of the UK aid programme, while a variety of agencies including DEFRA, DTI and other research councils focus on the UK, and particularly the improvement of its wealth generation. This is to be expected – after all the mandates of these various organizations are quite different. Yet as we have already seen with the issue of stakeholder participation there have been quite different experiences within each organization. DFID finds itself extolling the virtues of such participation throughout the developing world where it operates, while the home agencies of the country where DFID is based have had a far different thrust. The danger is that this may seem like 'do as I say not do as I do', and one can easily see how some in the developing world see this as a form of hypocrisy.

COMPROMISED PARTICIPATION IN OVERSEAS DEVELOPMENT

The participatory movement has had a much greater influence on agricultural research in the developing world than in Europe, although the impact that participatory methodologies per se have had on international development is difficult to assess. As alluded to in Chapter 5, some of the explanation for this lies with participation itself. There is distinction between participation as in participatory research methods and participation as in empowerment through those methods and a wider involvement of stakeholders in decision-making. Participatory research methods can be a very efficient means for collecting relevant and useful data under difficult and complex conditions. However, it should not be forgotten that the empowerment aspect of participation evolved partly as a response to the failures of earlier extension or delivery systems for getting farmers to adopt new technologies or approaches. Taking account of the social context and feeding it into the technology development and adoption should improve. Viewed thus, it could be argued that this approach is little more than a better method for technology delivery. This in essence leads one to conclude that in some instances far from being an empowering process it can be the opposite. Who organizes the workshops, designates the research area, sets the agenda and who attends? As pointed out by Cooke and Kothari (2001) participation obscures the relations of power and influence between elites and supposed beneficiaries. Thus, do participants participate out of free will? Furthermore, it is still the Western perception and concept of the best method put forward that is being pushed. Was it resource-poor farmers in the developing world that came up with the idea? Or was it our way of trying to incorporate what we think they wanted us to do? Just because it was a logical step away from systems that had inherent problems themselves, doesn't mean it was a step in the right direction. The crux of this is our seemingly endless search for the '$E=mc^2$' of human behaviour. Not only is this probably impossible, it also appears to be the based on the very same philosophical basis that proponents for participation purportedly abhor – mechanistic reductionism relying upon supposed experts for delivery.

That is not to say that participatory methods are useless or unnecessary in order to help identify the appropriate avenues for research and intervention. It is merely another approach that in some cases can help improve people's lives, and not in others. It is not a panacea.

Nonetheless, full and active participation by all stakeholders in project planning is expected by all international development sponsors, and in developing countries this does happen albeit to varying degrees. The extent, to which the final direction of research is influenced, is another matter.

COMPROMISED PARTICIPATION IN THE UK

It is interesting to contrast this with the UK systems. In the UK, it would be unthinkable to bring together a group of stakeholders such as farmers, representatives from industry, community groups, bankers, researchers, extension services, local government and research councils in order to discuss the establishment and form of a science-based research project. Anything resembling a PRA exercise to elicit stakeholder participation in setting a research agenda would probably be met by some amusement from those that manage research finances in the UK, and perhaps it is a little unrealistic to suggest it! There is a serious point here. Why should we see this process as patronizing here but fine for developing countries. Maybe it has something to do with the fact that participatory methods have been used here and questioned, not so much in relation to agricultural research alone, but as a management system. The difference is that it is known as human resource management and is under a different label. In the developing world you get participation here you get social psychology (Cooke and Kothari, 2001). And most of us have experienced the outcomes of participatory meetings or workshops that have little or nothing to do with what was actually said or even been on the agenda. Having said that, there is certainly a need to increase lines of communication between end users such as farmers and consumers, researchers and politicians in setting research priorities. This need is beginning to be recognized and advocated in publications such as the DTI Baker Report (1999), that provides support for researchers linking more closely with industrial and commercial end users, and the 'Science and Society' paper by the House of Lords regarding links with the general public as end users. However, in the case of the latter, informing or educating the public about what is done is one thing – giving them a real voice in setting agendas is something entirely different. It is this point more than any other, that the importance of power relationships must be realized. The influence of power relations in our agricultural research system, as previously stated, allowed farmers and via politicians the public, a greater role in the direction of research. Those power relations have shifted. Corporate power vested in agribusiness and large retailers now holds sway to the lament of the public and farmers alike. The BSE crisis, swine fever epidemic and debacle over GMOs all illustrate this point. With politicians of all persuasions espousing the merits of free trade and the increased power held by the WTO this is hardly surprising. This change has not gone unnoticed and as so eloquently related by Hertz (2001), this is not saying that private corporations do not have a role to play, nor that some of them have not been benevolent. After all, look at the impact that the Rockefeller Foundation has had in relation to agricultural development. What this shift does mean is that large corporations run by publicly unelected directors, answerable in the main to their shareholders, now have the power to make decisions and direct research for their own ends – profit. Furthermore, with deregulation and increased political clout these corporations are now probably less accountable. No wonder then, the public feels confused and disenfranchised over issues surrounding the production of our food. Before apportioning blame and heading for the next anti-capitalism demonstration, it is worth bearing in mind that under the CAP farmers were essentially only doing what society asked of them and that the research base reacted accordingly to meet the demands of many consumers, who at the same time as bemoaning the woes of modern agriculture were happily filling their baskets at the local supermarket. The public cannot have it both ways: we all bare some responsibility. Similarly, farmers blame supermarkets and the government for their economic dilemma; the government blames the farming industry; corporations blame the public; and the supermarkets pretend to just respond to public demand. If we do not like increased corporate power, decreased political power, GMOs, biotechnology and industrialized agricultural produce then we should at least attempt to sway our political leaders into a more proactive stance. An agricultural research policy placed firmly within the realms of a wider rural policy that more closely fits the needs and desires of as many stakeholders as possible with a higher level of public accountability would certainly be a move in the right direction. Sadly, this would appear to be some way off.

CONCLUDING COMMENTS

What has this 150-year-old potted history of agriculture told us? That most scientists don't work in a social vacuum and are not merely white-coated boffins stuck in laboratories doing 'their own thing'. Scientists practise science and by now it should have become apparent to the reader that science itself is more than a purely mechanistic approach to problem solving. More specifically it can be seen that science is an interaction between the supposed objectivity of scientific reasoning and the subjectivity of human thought processes. Thus it is always changing and adapting to the societal pressures that drive it. We have looked at these pressures and seen how they changed both temporally and spatially. This has indicated to us that in the UK, the research process has altered dramatically over this time with science playing an ever-increasingly dominant role in the development of agricultural technologies. Furthermore, that as power relations have shifted in determining which perceptions for the direction of research will be realized, agricultural scientists have shown a remarkable chameleon-like ability to adapt to those new realities. The results of this for the UK, have been to substantially increase the productivity of our farming systems, though not without many unforeseen environmental and human costs. The present situation is one where industrialized agriculture and corporate power hold sway. The science practised and the resultant technologies are a consequence of this. If there were to be a shift in emphasis resulting in alternative forms of agriculture production then research would drive and be driven by this. At present however, it appears that agricultural scientists and farmers are seeking solutions to today's problems mainly within the confines of current practices. As these practices apparently do not meet with the approval of large sections of our society, it is little wonder there is such discord. This in turn has led many to lose trust in science and the undoubted benefits it could provide if it were to work within the confines of a social context more equally representative of a greater proportion of our society's views, rather than as is perceived purely for the benefit of corporations. If politicians won't take a lead in representing these dissident voices and there is not enough consumer power to change things then it would appear that this situation will only alter in response to the economic impacts of increased free trade. One of the potential consequences or outcomes of this, which we have already discussed, could be an increase in the number of larger farmers and further vertical integration of the industry.

Overseas the situation though markedly different, does have many parallels in relation to how and why different approaches in agricultural research have evolved. In ex-colonial countries the purely extractive nature of the process led to a similar style of problem-solving often negating indigenous knowledge. Through political and economic changes the legacy of our colonial impact on developing world agriculture is still affecting how research is carried out and by whom. The establishment of participation as a methodology can be seen as a reaction to previous attempts at overcoming the constraints of agricultural development. Although some of the failings and weaknesses of this approach have been mentioned it is probably better than its predecessors. However, that does mean that it is not subject to the same issues of coercion, through the interplay of the power relationships of those involved. It is a pity that as this approach was starting to get established there was a parallel political shift, brought about to some extent, by the introduction of structural adjustment programmes and a subsequent loss of domestic autonomy over many areas of public spending and policy, including agriculture. One of the outcomes of this is that at the same time as participatory approaches may have been identifying areas that resource-poor farmers wanted addressing, the government would find itself in a position of being forced to reduce a particular agricultural subsidy thus potentially negating any positive impact the project might have had for its intended beneficiaries. The blatant hypocrisy of donor governments espousing the need to include participatory approaches in agricultural development programmes at the same time as supporting measures for introducing structural adjustment, beggars belief. With increased power residing in the hands of corporations, how compromised will participation become in setting agricultural research programmes. The introduction of GMOs and in particular a further loss of autonomy in relation to gene patenting are obvious concerns here.

At a macro level in both the UK and in developing countries it has been demonstrated that participation is always compromised by those in a position to do so and the outcome of these compromises is a complex interaction involving the power relationships of all of those involved. What is not clear at present is what these outcomes will be. In the UK, Monsanto had its fingers badly burned when it misread public concern over GMOs. To what extent it has taken on board these concerns is yet to be seen. The situation with gene patenting and the impact on developing country agricultural development is also unclear.

Agricultural scientists will most probably continue to work within the disciplinary limits set by their peers within the context of their particular institutions governed by the complex interweaving of the ever-changing societal pressures that dictate the value, direction and funding for their research. Agricultural research will continually vex us as with the nature of anything so innately and humanly complex we sometimes lose sight of what it is we set out to do – feed ourselves.

Perhaps TS Eliot was right when he said: 'We shall not cease from exploration and the end of all our exploring will be to arrive where we started and know the place for the first time' (*Gerontion*).

References

ACARD (1992) *Comprehensive Survey of Research and Technology Foresight in the United Kingdom*, commissioned by the Office of Science and Technology 1992, HMSO, London

Agricultural Research 1931–1981, GW Cooke (ed), ARC, London, 1981

Audi, P and Mills, B (1998) 'Translating farmer constraints into research themes', in B Mills (ed), *Agricultural Research Priority Setting Information Investments for the Improved Use of Research Resources*, pp41–50, ISNAR, The Hague

Baker, J (1999) *Creating Knowledge, Creating Wealth*, report to the Minister of Science and the Financial Secretary to the Treasury, August 1999

Baldwin, KDS (1957) *The Niger Agricultural Project: An Experiment in African Development*, Basil Blackwell, Oxford

BBSRC (Mission Statement) – taken from the BBSRC website at www.bbsrc.ac.uk

Becker, T (2000) 'Participatory research in the CGIAR', a discussion paper prepared for the NGO-workshop *Food For All – Farmer First in Research*, accompanying the GFAR 2000 in Dresden

Bell, S and Morse, S (1999) *Measuring the Immeasurable?*, Earthscan, London

Benor, D and Harrison, JQ (1977) *Agricultural Extension: The Training and Visit System*, World Bank, Washington

Bevan, P (2000) 'Who's a goody? Demythologizing the PRA agenda', *Journal of International Development*, Vol 12, pp751–759

Biggs, SD (1989) 'Resource-poor farmer participation in research, a synthesis of experiences from nine national agricultural research systems', *OFCOR Comparative Study Paper*, No 3, ISNAR, The Hague

Biggs, S and Matsaert, H (1998) 'An actor orientated approach for strengthening research and development capabilities in natural resource systems', *Public Administration and Development*, Vol 19, No 3, pp231–262

Biggs, S and Smith, G (1997) 'Beyond methodologies, coalition-building for participatory technology development, *World Development*, Vol 26, No 2, pp239–248

Biggs, S and Smith, G (1998) 'Contending coalitions in agricultural research and development, challenges for planning and management', *Knowledge and Policy*, Vol 10, No 4, pp77–89

Blackburn, J and Holland, J (eds) (1998a) *Who Changes? Institutionalizing Participation in Development*, IT Publications, London

Blackburn, J and Holland, J (1998b) 'General introduction', in Blackburn, J and Holland, J (eds), *Who Changes? Institutionalizing Participation in Development*, IT Publications, London, pp1–8

Boserup, E (1965) *The Conditions of Agricultural Growth: The Economics of Agrarian Change under Population Pressure*, Allen and Unwin, London

Burns, A (1948) *History of Nigeria*, (4th Edition), Allen and Unwin, London

Carney, D (ed) (1998) *Sustainable Rural Livelihoods, What Contribution Can We Make?* DFID, London

Carson, R (1962) *Silent Spring*, Fawcett, New York

Chambers, R (1983) *Rural Development: Putting the Last First*, Longman, London

Chambers, R (1992) *Rural Appraisal: Rapid, Relaxed and Participatory*, Discussion paper 311, IDS, University of Sussex, Brighton

Chambers, R (1993) *Challenging the Professionals: Frontiers for Rural Development*, Intermediate Technology, London

Chambers, R (1997) *Whose Reality Counts? Putting the Last First*, Intermediate Technology, London

Chambers, R and Ghildydal, BP (1985) *Agricultural Research for Resource-Poor Farmers: The Farmer-First-and-Last Model*, Discussion paper 203, IDS, University of Sussex, Brighton

Chambers, R and Jiggins, J (1986) *Agricultural Research for Resource-Poor Farmers: A Parsimonious Paradigm*, Discussion paper 220, IDS, University of Sussex, Brighton

Chambers, R, Pacey, A and Thrup, LA (eds), (1989) *Farmer First: Farmer Innovation and Agricultural Research*, Intermediate Technology, London

Clarke, J (1981) 'Households and the political economy of small-scale cash crop production in south-western Nigeria', *Africa*, 51 (4) pp807–823

Cleaver, HM (1972) 'The contradictions of the Green Revolution', *The American Economic Review*, 72, pp177–188

Commission of the European Communities (1991) *Reforming the Common Agricultural Policy*, Background Report, ISEC/B8/91, Storey's Gate, London

Conway, G R (1986) *Agroecosystem Analysis for Research and Development*, Winrock International, Bangkok

Conway, G (1997) *The Doubly Green Revolution*, Penguin, London

Cooke, B and Kothari, U (eds), (2001) *Participation: The New Tyranny*, Zed Books, London

Cox, A, Farrington, J and Gilling, J (1998) *Reaching the Poor? Developing a Poverty Screen for Agricultural Research Proposals*, ODI Working Paper 112 ODI, London

Cox, D, Georghion, L and Salazar, A (2000) *Links to the Science Base of the Information Technology and Biotechnology Industries*, a report on behalf of the ESRC for the Director General of Research Councils, PREST, Manchester

Cunningham, P (1998) *Science and Technology in the United Kingdom*, 2nd edition Cartermill International, London

Curtis, J (1860) *Farm Insects: Being the Natural History and Economy of the Insects Injurious to the Field Crops of Great Britain and Ireland*, Blackie and Son, London

DFID (1997) *Eliminating World Poverty: A Challenge for the 21st Century*, DFID, London

DFID (2000) *Eliminating World Poverty: Making Globalisation Work for the Poor*, Stationery Office, Norwich

DTI (2001) Report on the Research Assessment Exercise 2001 at www.dti.gov.uk/rae2001/intro.htm

Ellis, F (1998) 'Livelihood diversification and sustainable rural livelihoods', in Carney, D, *Sustainable Rural Livelihoods: What Contribution can We Make?*, paper presented for DFID, natural resources advisors conference, DFID, London, July 1998, pp53–65

Eponou, T (1993) *Partners in Agricultural Technology. Linking Research and Technology Transfer to Serve Farmers*, ISNAR, The Hague

Estrella, M and Gaventa, J (1997) *Who Counts Reality? Participatory Monitoring and Evaluation: a Literature Review*, IDS Working Paper 70, IDS, University of Sussex, Brighton

European Commission (2000) *Towards a European Research Area*, Communication from the Commission to the Council, The European Parliament, The Economic and Social Committee and the Committee of the Regions, The Office of Official Publication of the European Communities, Luxembourg

Falcon, W P (1970) 'The Green Revolution: generations of problems', *American Journal of Agricultural Economics*, 52, pp698–710

Farmer, B H (1986) 'Perspectives on the 'Green Revolution' in South Asia' *Modern Asian Studies*, 20 (1), pp175–199

Faulkner, O T and Mackie, J R (1933) *West African Agriculture*, Cambridge University Press, Cambridge

Financial Times (1988) EEC 'Fruit waste is bananas', *Financial Times*, London, 18 February

Fletcher, TW (1961) 'The Great Depression of English Agriculture 1873–1896', *Economic History Review* 2nd Series, Vol 13, No 3, pp417–432

Foresight 'Foresight – making the future work for you' – taken from the Foresight website: http://www.foresssight.gov.uk/

Forrest, T (1981) 'Agricultural policies in Nigeria 1900–78', in *Rural Development in Tropical Africa*, Heyer, J, Roberts, P and Williams, G (eds), MacMillan, Basingstoke, pp222–251

Forrest, T (1995) *Politics and Economic Development in Nigeria*, updated edition, Westview Press, Boulder

Francis, P (1984) 'For the use and common benefit of all Nigerians: consequences of the 1978 land nationalization', *Africa*, 54 (3), pp5–27

Freire, P (1972) *Cultural Action for Freedom*, Penguin, London

Garforth, C (1997) 'Institutionalizing innovations: a singular concern for R&D', AERDD Working Paper 97/1, Reading University

Glaeser, B (ed), (1987) *The Green Revolution Revisited, Critique and Alternatives*, Allen and Unwin, London

Goddard, N and RASE (Royal Agricultural Society of England) (1988) *Harvests of Change: The Royal Agricultural Society of England 1838–1988*, Quiller, PR, London

Goldman, A and Smith, J (1995) 'Agricultural transformations in India and northern Nigeria: exploring the nature of Green Revolutions', *World Development*, 23 (2), pp243–263

Goodell, G (1984) 'Bugs, bunds and bottlenecks: organizational contradictions in the new rice technology', *Economic Development and Cultural Change*, 33, pp23–41

Goodman, D, Sorj, B and Wilkinson, J (1987) *From Farming to Biotechnology: A Theory of Agro-industrial Development*, Basil Blackwell, Oxford

Gras, NSB (1940) *A History of Agriculture in Europe and America*, F S Crafts and Co, New York

Grigg, D (1989) *English Agriculture*, Blackwell Publishers, Oxford

Guyer, J (1993) 'Toiling ingenuity: food regulation in Britain and Nigeria', *American Ethnologist*, 20 (4), pp797–817

Hambly, H and Setshwaelo, L (1997) *Agricultural Research Plans in Sub-Saharan Africa*, Research report 11, ISNAR, The Hague

Harrison, P (1987) *The Greening of Africa: Breaking Through in the Battle for Land and People*, Paladin Gritton, London

Hart, K (1982) *The Political Economy of West African Agriculture*, Cambridge University Press, Cambridge

Hartmans, E H (1997) 'Some issues and priorities for the CGIAR in global agricultural research', in Bonte-Friedheim, C and Sheridan, K (eds), *The Globalization of Science: the Place of Agricultural Research*, ISNAR, The Hague, pp47–51

Harvey, J and Potten, DH (1987) 'Rapid Rural Appraisal of small irrigation schemes in Zimbabwe', *Agricultural Administration and Extension*, 27, pp141–155

Helleiner, GK (1966) *Peasant Agriculture, Government, and Economic Growth in Nigeria*, Richard D Irwin Inc, Homewood

Hertz, N (2001) *The Silent Takeover Global Capitalism and the Death of Democracy*, William Heinemann, London, HMSO website at www.hmso.gov.uk

HMSO (1993) *Comprehensive Survey of Research and Technology Foresight in the United Kingdom*, commissioned by the office of Science and Technology in 1992

HMSO (1993) *Comprehensive Survey of Research and Technology Foresight in the United Kingdom*, Office of Science and Technology 1192, London

Holland, J and Blackburn, J (1998) *Whose Voice? Participatory Research and Policy Change*, IT Publications, London

Howells J, Nedeva M and Georghiou L (1998) *Industry–Academic Links in the UK*, PREST, University of Manchester, Manchester

Hughill, A (1978) *Sugar and All That...A History of Tate and Lyle*, Gentry

Hutchinson, J and Owens, AC (1980) *Change and Innovation in Norfolk Farming*, Packard, Chester

IDS (1996) *The Power of Participation: PRA and Policy*, Policy Briefing Issue 7, IDS, Brighton

IDS (1998) *Participatory Monitoring and Evaluation: Learning from Change*, Policy Briefing Issue 12, IDS, Brighton

Idachaba, F (1998) *Instability of National Agricultural Research Systems in Sub-Saharan Africa: Lessons from Nigeria*, ISNAR Research Report 13, ISNAR, The Hague

IITA (1988) *IITA Strategic Plan 1989–2000*, IITA, Ibadan

Joly P-B (2000) 'Contested innovation, what lessons may be drawn from public controversies on GMOs? Proc III Int Symp on Brassicas', GJ King (ed), *Acta Horticultura*, 539, pp33–38

Jones, G H (1936) *The Earth Goddess: A Study of Native Farming on the West African Coast*, Longman, Green and Co, London

Kumar, K (ed) (1993) *Rapid Appraisal Methods*, World Bank, Washington

Klug, A (2000) Presidential address, Royal Society media release, www.royalsoc.ac.uk

Levins, R (1974) 'The qualitative analysis of partially specified systems', *Annals of the New York Acadamy of Science*, 231, pp123–138

Lipton, M (1988) 'The place of agricultural research in the development of sub-Saharan Africa', *World Development*, 16 (10), pp1231–1257

Lipton, M (1989) *New Seeds and Poor People*, Unwin Hyman, London

Loveridge D, Georghiou L and Nedeva M (1995) 'United Kingdom Technology Foresight Programme – Delphi Survey', PREST, University of Manchester, Manchester

MAFF (1931) *Report on the Sugar Beet Industry at Home and Abroad*, HMSO, London

Maredia, M, Howard, J and Boughton, D (1997) *No Shortcuts to Progress: An Assessment of Agricultural Research Planning and Priority Setting in Africa*, Policy Synthesis for USAID, Bureau for Africa Office of Sustainable Development, no 29

Martin, S (1993) 'From agricultural growth to stagnation: the case of the Ngwa, Nigeria, 1900–1980', in *Population Growth and Agricultural Change in Africa*, Turner, BL, Hyden, G and Kates, RW (eds), University Press of Florida, Gainesville, pp302–323

Masefield, G B (1972) *A History of the Colonial Agricultural Service*, Clarendon Press, Oxford

Mbabu, A, Mills, B and Lynam, J (1998) 'Information and human resource investments for research priority setting', in *Agricultural Research Priority Setting. Information Investments for the Improved Use of Research Resources*, Mills, B (ed), ISNAR, The Hague, pp87–98

McCracken, J A (1988) 'A working framework for rapid rural appraisal: lessons from a Fiji experience', *Agricultural Administration and Extension*, 29, pp163–184

Meagher, K (1990) 'Institutionalising the bio-revolution: implications for Nigerian smallholders', *Journal of Peasant Studies*, 18 (1), pp68–89

Monyo, J (1997) 'The plight of national agricultural research systems in low-income, food-deficit countries', in Bonte-Friedheim, C and Sheridan, K (eds), *The Globalization of Science: The Place of Agricultural Research*, ISNAR, The Hague, pp131–137

Morse, S, McNamara, N, Acholo, M and Okwoli, B (2000) *Visions of Sustainability. Stakeholders, Change and Indicators*, Ashgate, Aldershot

Narayan, D (1997) *Voices of the Poor: Poverty and Social Capital in Tanzania*, World Bank, Washington DC

Newby, H (1988) *Country Life – A Social History of Rural England*, Cardinal, London

Nickel, J L (1997) 'A global agricultural research system for the 21st century', in Bonte-Friedheim, C and Sheridan, K (eds), *The Globalization of Science: the Place of Agricultural Research*, ISNAR, The Hague, pp139–155

Norton, A (1998) 'Analysing participatory research for policy change', in Holland, J and Blackburn, J, *Whose Voice? Participatory Research and Policy Change*, IT Publications, London, pp179–191

Oehmke, J, Anandajaysaekeram, P and Masters, W (1997) *Agricultural Technology Development and Transfer in Africa. Impacts Achieved and Lessons Learned*, technical paper no 77, SD Publication Series, Office of Sustainable Development Bureau for Africa

Okali, C, Sumberg, J and Farrington, J (1994) *Farmer Participatory Research*, Intermediate Technology, London

OST (1993) *Realising our Potential, a Strategy for Science, Engineering and Technology*, Cmnd 2250, HMSO, London

Oyenuga, V (1967) *Agriculture in Nigeria An Introduction*, FAO, Rome

Pardey, P, Roseboom, J and Beintema, N (1995) *Agricultural Research in Africa: Three Decades of Development*, ISNAR briefing paper 19, ISNAR, The Hague

Porter, RW and Prysor-Jones, S (1997) *Making a Difference to Policies and Programs: A Guide for Researchers*, Support for Analysis and Research in Africa Project, Washington DC

Pretty, J (1995) *Regenerating Agriculture: Policies and Practice for Sustainability and Self-reliance*, Earthscan, London

Pretty, J (1998) *Participatory Learning for Integrated Farming*, in Proceedings of the internet conference on integrated bio-systems, Eng-Leong Foo and Della Senta, T (eds), www.ias.unu.edu/proceedings/icibs/jules/paper.htm

Richards, P (1983) 'Ecological change and the politics of African land use', *African Studies Review*, 26 (2), pp1–72

Richards, P (1985) *Indigenous Agricultural Revolution*, Unwin Hyman, London

Robb, C (1998) 'PPAs: a review of the World Bank's experience', in Holland, J and Blackburn, J, *Whose Voice? Participatory Research and Policy Change*, IT Publications, London, pp131–142

Rothschild (1953) 'Agricultural research 1953', published in the Jubilee Volume of the Long Ashton Research Station, *Science and Fruit*, University of Bristol, Bristol

Rothschild (1971) 'The organization and management of government R&D – the Rothschild report', included in the Green Paper (Cmnd 4814) *A Framework for Government Research and Development*

Roy, S (1990) *Agriculture and Technology in Developing Countries: India and Nigeria*, Sage Publications India, New Delhi

Ruttan, V (1977) 'The Green Revolution: seven generalizations', *International Development Review*, 4, pp16–23

Sampson, H and Crowther, E (1943) 'Crop production and soil fertility problems', *The West Africa Commission 1938–39 Technical Reports (Part 1)*, Leverhulme Trust, London

Sellamna, N-E, (1999) *Relativism in Agricultural Research and Development: Is Participation a Post-Modern Concept?*, ODI Working Paper 119, Overseas Development Institute, London

Sen, A (1981) *Poverty and Famines*, Clarendon Press, London

Shiva, V (1991) 'The Green Revolution in the Punjab', *The Ecologist*, 21 (2), pp57–60

Spedding C (1984) 'Agricultural research policy', in Goldsmith, M (ed), *UK Science Policy: a Critical Review of Policies for Publicly Funded Research*, Longman, London

Tracey, M (1989) *Government and Agriculture in Western Europe 1880–1988*, Harvester Wheatsheaf, London

Toulmin, C and Chambers, R (1990) *Farmer-First: Achieving Sustainable Dryland Development in Africa*, Issues paper/Drylands Programme no 19, IIED, London

Tribe, D (1997) 'The best-kept secret', in Bonte-Friedheim, C and Sheridan, K (eds), *The Globalization of Science: The Place of Agricultural Research*, ISNAR, The Hague, pp195–202

Trigo, E (1997) 'The role of NARS in the changing global agricultural research system', in Bonte-Friedheim, C and Sheridan, K (eds), *The Globalization of Science: The Place of Agricultural Research*, ISNAR, The Hague, pp204–209

Walker, B, Norton, G, Conway, G, Comins, H and Birley, M (1978) 'A procedure for multidisciplinary ecosystem research: with reference to the South African Savanna Ecosystem Project', *Journal of Applied Ecology*, 15, pp481–502

Wallace, T and Martin, J (1954) *Insecticides and Colonial Agricultural Development*, Butterworths Scientific Publications, London

Wambugu, F (2000) 'Feeding Africa opinion interview', *New Scientist*, 27 May, pp41–43

Watts, H (1970) 'The location of the sugar beet industry in England and Wales', *Transactions of the British Institute of Geographers*, 152

Williams, G (1988) 'Why is there no Agrarian capitalism in Nigeria?', *Journal of Historical Sociology*, 1 (4), pp345–398

Williams, M H and Stevenson, J (1999) (eds) *Observations of Weather: The Weather Diary of Sir John Wittewronge of Rothamsted, 1684–1689*, Hertforshire, Record Society, Hitchen

World Bank (2000a) *What is the Impact of Agricultural Research in Africa?*, www.worldbank.org/html/aftsr/impact1.htm

World Bank (2000b) *How to Make Agricultural Research Perform Better in Africa*, www.worldbank.org/html/aftsr/reform1.htm

World Bank (2000c) *Sustainable Financing Initiative*, www.worldbank.org/html/aftsr/sfwhat1.htm

World Bank (2001) *The World Development Report 2000–2001. Attacking Poverty*, Oxford University Press, Oxford

Yudelman, M (1997) 'Agricultural research in the tropics: past and future', in Bonte-Friedheim, C and Sheridan, K (eds), *The Globalization of Science: The Place of Agricultural Research*, ISNAR, The Hague, pp211–216

Index